W9-AGF-323

A Practical Guide to Viewing the Universe

TERENCE DICKINSON

Foreword by Timothy Ferris

ILLUSTRATIONS BY

ADOLF SCHALLER, VICTOR COSTANZO AND ROBERTA COOKE

PRINCIPAL PHOTOGRAPHY BY TERENCE DICKINSON

FIREFLY BOOKS

Published by Firefly Books Ltd. 1998
© 1998 Terence Dickinson

All rights reserved. No part of this publication may be reproduced, stored in a retrieval system or transmitted in any form or by any means, electronic, mechanical, photocopying, recording or otherwise, without the prior written consent of the Publisher.

Reprinted with revisions December 1998; April 1999

Library of Congress Cataloging-in-Publication Data is available.

Canadian Cataloguing in Publication Data

Dickinson, Terence
NightWatch : a practical guide to viewing the universe

Rev. ed.
Includes bibliographical references and index.
ISBN 1-55209-300-X (bound)
ISBN 1-55209-302-6 (pbk.)

1. Astronomy—Popular works. I. Title.

QB64.D52 1998 520 C98-931005-1

Published in Canada in 1998 by
Firefly Books Ltd.
3680 Victoria Park Avenue
Willowdale, Ontario M2H 3K1

Published in the United States in 1998 by
Firefly Books (U.S.) Inc.
P.O. Box 1338, Ellicott Station
Buffalo, New York 14205

Produced by
Bookmakers Press Inc.
12 Pine Street
Kingston, Ontario K7K 1W1
(613) 549-4347
tcread@sympatico.ca

Design by
Roberta Cooke

Color separations by Friesens, Altona, Manitoba

Printed and bound in Canada by Friesens, Altona, Manitoba

Printed on acid-free paper

The Publisher acknowledges the financial support of the Government of Canada through the Book Publishing Industry Development Program for its publishing activities.

Other Firefly books by Terence Dickinson

Exploring the Night Sky
Exploring the Sky by Day
The Universe and Beyond
The Backyard Astronomer's Guide (with Alan Dyer)
From the Big Bang to Planet X
Extraterrestrials (with Adolf Schaller)
Other Worlds
Summer Stargazing
Splendors of the Universe (with Jack Newton)

CREDITS AND ACKNOWLEDGMENTS

Photographs in this book that are not credited below are by Terence Dickinson.

p.4 left, Bernard Clark; p.7, Jim Riffle; p.9, George Greaney; p.13, NASA; p.14 bottom left, NASA; p.14-17, illustrations by Adolf Schaller; p.15 upper left, NASA; p.16 upper left, NASA; p.17 upper right, Jack Newton; p.18 upper left, NASA; p.18 lower left, illustration by John Bianchi; p.18 right, illustration by Adolf Schaller; p.19, Space Telescope Science Institute; p.20-22, illustrations by Adolf Schaller; p.23 top (both) Jack Newton; p.23 bottom, NOAO; p.27, Jerry Lodriguss; p.28-29, illustrations by Victor Costanzo; p.29 right, Alan Dyer; p.30-32, illustrations by Roberta Cooke; p.35 inset, NASA; p.39, Alan Dyer; p.43 bottom, Jerry Lodriguss; p.44, illustration by Victor Costanzo; p.45, illustration by Roberta Cooke; p.50, illustration by Victor Costanzo; p.51, illustration by Roberta Cooke; p.54, illustration by Victor Costanzo; p.55, illustration by Roberta Cooke; p.56, Jerry Lodriguss; p.58, illustration by Victor Costanzo; p.59, illustration by Roberta Cooke; p.62 lower left, Alan Dyer; p.67, illustration by Victor Costanzo; p.70 bottom left and right, Alan Dyer; p.72 left, Alan Dyer; p.72 right, Meade Instruments; p.74 top, Meade Instruments; p.74 center, Gary Collver; p.74 bottom, Alan Dyer; p.65, Alan Dyer; p.77 top right, Alan Dyer; p.79 center and bottom (both), Alan Dyer; p.80 top and #3, Alan Dyer; p.80 bottom, Meade Instruments; p.84 inset, Alan Dyer; p.85 top, Tony Hallas; p.87-90, Jerry Lodriguss; p.91 top and center, illustrations by Adolf Schaller; p.91 right, Jerry Lodriguss; p.92 left (both), illustrations by Adolf Schaller; p.92 right, Jack Newton; p.93 top right, center and bottom, illustrations by Adolf Schaller; p.93 top right, Jack Newton; p.94 illustration by Adolf Schaller; p.95 *Sky Atlas 2000* segment courtesy Sky Publishing; p.95 right, Jerry Lodriguss; p.96 top right, illustration by Adolf Schaller; p.96, notebook courtesy Russell Sampson; p.98 top, illustration by Adolf Schaller; p.98 #2, Jerry Lodriguss; p.98 #3, Jack Newton; p.100-119, cartography by Roberta Cooke from base photography by Ray Villard; p.120, Robert May; p.121, Space Telescope Science Institute; p.122, illustration by Victor Costanzo; p.124 top, George Liv; p.125 top (both), Roy Bishop; p.126, NASA; p.127, NASA; p.127 inset, Frank Hitchens; p.130, NASA; p.131, notebook by Matthew Sinacola; p.132 left, notebook courtesy Russell Sampson; p.132 right, Paul Doherty; p.139 left, sketch by Matthew Sinacola; p.142 top, Wolfgang Lille; p.144 bottom, John Hicks; p.144 (both), Alan Dyer; p.145 left, Alan Dyer; p.145 top, Dan Falk; p.145 bottom, John Nemy; p.146, Ron Schmidli; p.148, Andreas Gada; p.149 right, Ron Schmidli; p.150 top, Alan Dyer; p.150 left center, John Nemy; p.150 right, illustration by Roberta Cooke; p.151 left, Alan Dyer; p.151 right, Dan Falk; p.152 top, Jim Failes; p.157 top, William Broderick; p.158 top, Cathy Hall; p.158 left, STScI; p.158 bottom, Alan Dyer; p.160 left, illustration by Roberta Cooke; p.160 top, illustration by Victor Costanzo; p.161 right, John Mirtle; p.162 left, Michel Tournay; p.162 bottom, Bev McConnell; p.167 top left, Alan Dyer; p.167 #2, John Nemy; p.169 center, Meade Instruments; p.169 bottom, Jack Newton; p.171, Calgary Planetarium.

Sometimes "revised edition" means tacking on a few paragraphs here and there to bring a book up to date. Not this time. Every page has been revised, redesigned and augmented by loads of new photographs and illustrations. Graphic designer Roberta Cooke has brilliantly created a completely new look for ***NightWatch*** yet maintained all the popular features of the previous editions. Of many colleagues who kindly offered photographs and advice, I especially appreciate the contributions of Alan Dyer and Jerry Lodriguss. Christine Kulyk commented on the entire manuscript and was particularly helpful with the revision of the 20-chart star atlas. A tip of the hat to proofreaders Catherine Delury and Charlotte DuChene. And, finally, a big hug and thank you to my wife Susan, production manager and chief copy editor. I couldn't have done it without you.

To Susan, who continues to share the celestial voyage

CONTENTS

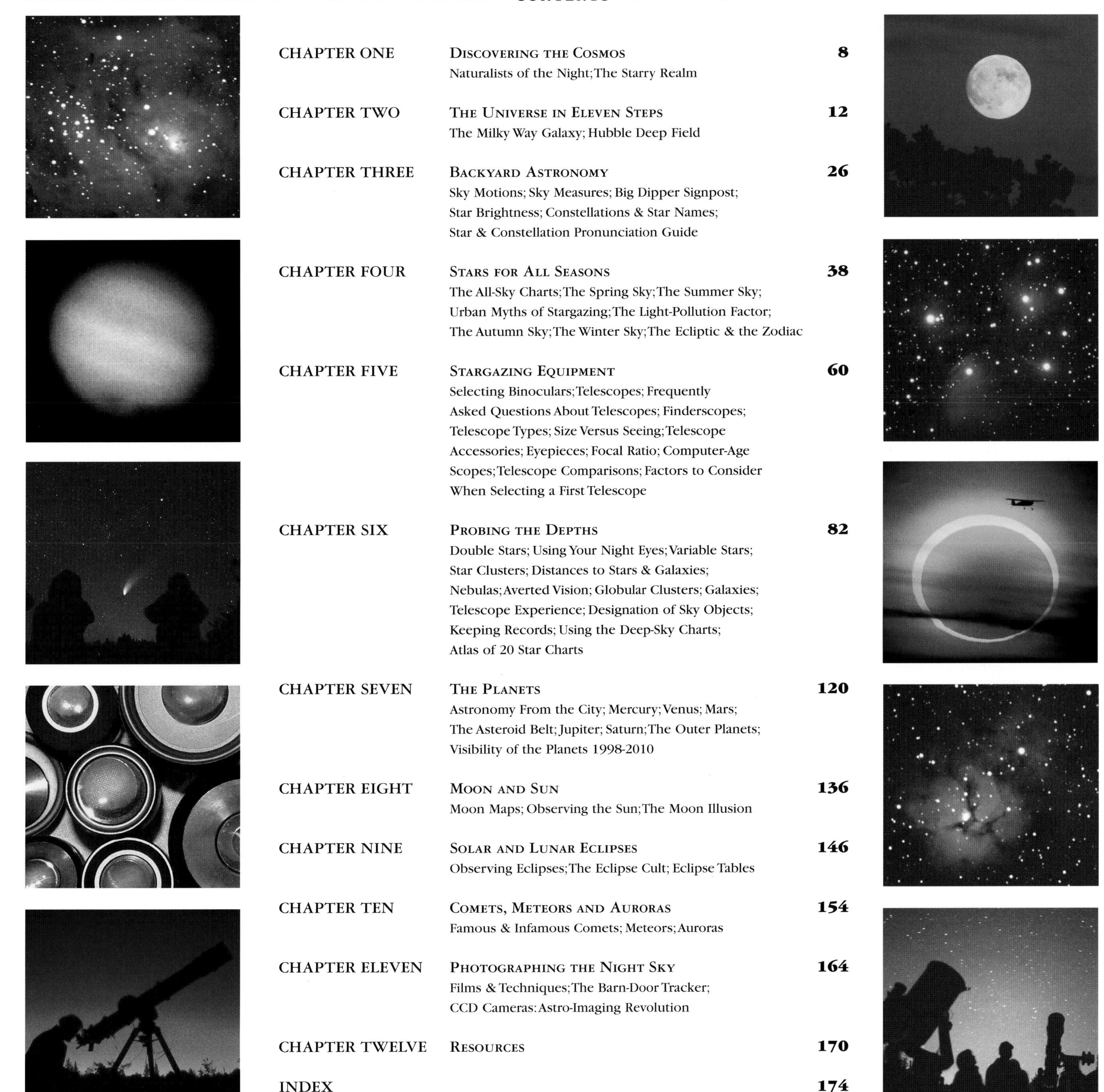

Terence Dickinson is the author of 14 astronomy books, the editor of ***SkyNews*** magazine and an astronomy columnist for ***The Toronto Star*** and the Canadian Discovery Channel. His lifelong interest in astronomy began at age 5 when he saw a brilliant meteor from the sidewalk in front of his home. In the 1960s and 1970s, he was a staff astronomer at two major planetariums and, since 1976, has been a full-time astronomy writer and editor. He has received numerous national and international awards for his work, among them the New York Academy of Sciences book of the year award and the Astronomical Society of the Pacific's Klumpke-Roberts Award for outstanding contributions in communicating astronomy to the public. In 1994, asteroid 5272 Dickinson was named after him. In 1995, he received the Order of Canada, the nation's highest civilian achievement award. He is also an accomplished astrophotographer, having taken most of the pictures in this book himself (e.g., photograph at right, taken from his backyard). He lives in rural eastern Ontario where he enjoys dark night skies above his backyard roll-off roof observatory crammed with telescopes and astrocameras.

The approach of the next millennium finds growing numbers of humans taking an interest in our planet's cosmic environment, not just through reading but also by going out at night to have a look for themselves. Stargazing is a highly rewarding activity. I've yet to meet anyone who regretted having taken the trouble to learn his or her way around the night sky, and as with other pursuits, the more one knows about what one is seeing, the more fulfilling it becomes.

Which is why it's important to have the right book, a stargazing companion that awakens, informs and keeps you reading. Like a good human companion, it should be knowledgeable and reliable but also concise, conveying the essentials without interposing itself between you and the stars. It should be robust and easy to use, whether you're observing eclipses of the moons of Jupiter through an 8-inch telescope, scanning the Milky Way through binoculars or just out learning a few constellations and star names.

Most veteran observers have a few favorite books that have survived the twin tests of field use and indoor scrutiny to become old friends—tattered, dew-soaked troopers that have won their loyalty over countless days and nights. Among mine are H.A. Rey's *The Stars: A New Way to See Them*, an eccentric but timelessly charming introduction to the constellations; Allan Sandage's *Hubble Atlas of Galaxies*, which resembles what Christopher Columbus might have published had he been a better writer—and been able to take a camera along; and Robert Burnham Jr.'s *Burnham's Celestial Handbook*, a labor of love that, though inadequately requited while Burnham was alive, will long preserve his memory.

Terence Dickinson's *NightWatch* shows real promise of becoming such a book. In the years since it was first published, it has earned the trust and affection of a widening circle of amateur astronomers on all levels. The reasons are not difficult to discern. Dickinson is himself both a skilled observer and a lucid writer. He knows what's out there and how best to see it, and he shares his expertise in the spare, friendly voice of someone who has educated not only himself but many others. His deep aesthetic appreciation of astronomy is reflected in the book's splendid charts and illustrations. He has kept refining and updating *NightWatch* until it fits into an observer's routine as comfortably as a sharp old knife in an angler's hand. Like a good night sky, *NightWatch* is clear and wind-free. Try it and see for yourself.

—Timothy Ferris, Emeritus Professor
University of California, Berkeley

In the years since the first edition of *NightWatch* appeared in 1983, more than a quarter of a million copies have found their way into the hands of astronomy enthusiasts in Canada and the United States. For me, the most gratifying aspect of this successful publishing story is the feedback from so many backyard astronomers who say that the book was their primary guide during the crucial initial stages of their celestial explorations.

As in the previous editions, the overriding goal in this revised and expanded version has been to provide a *complete* first book of amateur astronomy. I wanted to retain the features that readers say they like, so I have not tampered with the basic structure and presentation. But extensive fine-tuning and updating have touched almost every page. The most visible of the changes is that the book is thicker (16 more pages), but there is much more than that. Most of the photographs have been replaced with more relevant—or simply superior—images, and an improved layout by ace graphic designer Roberta Cooke has allowed more cogent organization of the material. The 20 atlas charts have been reworked to include additional objects of interest as well as new, more accurate Hipparcos satellite distances for over 250 stars down to magnitude 3.5.

Other changes include almost total rewrites and expansions of several sections to reflect the ongoing evolution of amateur telescope equipment and accessories. Tables have been updated throughout. Many other improvements are the direct result of suggestions from readers of previous editions. One feature that has not changed, however, is the coil binding, which many readers have told me they prefer for its durability and ease of use outdoors. As before, prices throughout the book are in U.S. dollars.

Although more people than ever before are dabbling in recreational astronomy and the range and quality of equipment to pursue the hobby have never been better, an ever-growing menace is threatening the night we cherish. The foe is light pollution—the glare spilled from streetlamps, outdoor-sign illumination, parking-lot lights, building security lights and outdoor fixtures around private residences. Any one of these sources can ruin your backyard view of the night sky. Even if your observing site is protected from direct interference, outdoor lighting in general produces giant glowing domes over our cities and towns that have beaten back the stars.

Because the glow is visibly growing every year, those who wish to see the natural beauty of a dark night sky must flee ever farther into the country. For many aficionados, an evening of stargazing has become an expedition. But all is not gloom and doom. The dark cloud cast by light pollution has turned out to have an intriguing silver lining. Far from diminishing interest in astronomy, urban sky glow seems to have fueled it. When our grandparents were young, a view of the night sky strewn with stars and wrapped in the silky ribbon of the Milky Way could be seen from the front porch. Today, for most people, it is a relatively rare and exotic sight, something to be talked about and cherished as a memory.

Many family vacations now include plans for dark-site stargazing. Each year, thousands of astronomy enthusiasts gather at conventions and summer "star parties" far from city lights to share their interest. As urban glow inexorably marches deeper into the countryside, the 21st century will see the emergence of dark-sky preserves—areas intentionally set aside in state, provincial or national parks where there are no obtrusive lights and never will be. These shrines to the glory of the starry night will become ever more precious in the decades ahead.

Terence Dickinson
Yarker, Ontario
July 1998

The Orion Nebula is suspended like a colossal cosmic flower tossed into the blackness of space and frozen in time. This is just one of nature's many celestial treasures that await the curious gaze of Earthlings who take the challenge of personal exploration of the starry night sky.

CHAPTER ONE

DISCOVERING THE COSMOS

We are voyagers on the Earth through space, as passengers on a ship, and many of us have never thought of any part of the vessel but the cabin where we are quartered.

S.P. Langley

Imagine a world where a thimbleful of matter weighs as much as Mount Everest. The gravitational pull at the surface of this celestial body is so enormous that a human visitor would be crushed instantly by his or her own weight into a puddle no thicker than an atomic nucleus.

Now consider a planet whose sky is ablaze with the fireworks of stars being ripped into gaseous tendrils by a bizarre gravity funnel. There can be no living creatures in this corner of the universe, for the region is perpetually bathed in ultralethal doses of x-ray radiation.

Envision a planet cloaked in a choking blanket of carbon dioxide laced with sulfuric-acid rain. Here, it is so hot that lead would be a convenient liquid to use in a thermometer. A human explorer would be simultaneously incinerated and asphyxiated in the hellish environment.

Then there is a realm where two suns illuminate the cosmic landscape. The solar twins waltz in a deadly gravitational embrace, with millions of tons of gaseous star-stuff flowing between them every second. The tug-of-war will end in the explosive destruction of one of the stars, reducing any nearby worlds to cinders.

All these alien vistas exist. The dense supergravity object is the collapsed core of an exploded star, known as a neutron star. The gravity funnel is a massive black hole that is believed to be lurking at the center of the Milky Way Galaxy. The carbon-dioxide hothouse is Venus, the nearest planet to Earth and the brightest object in the night sky after the Moon. The twin-sun system is known as Beta in the constellation Lyra, seen overhead on midsummer evenings. In a universe of a billion trillion stars and unknown trillions of planets, the imagination withers in any attempt to grasp the true diversity of the cosmos.

Yet for me, as I stand under the nighttime canopy of stars, *that* has always been the lure of the backyard exploration of the

A lifetime of wonder and the satisfaction of personally discovering the universe await anyone who chooses to know the night sky.

Left: Astronomy enthusiasts enjoy a night of stargazing. Above: The Andromeda Galaxy is a vast disk-shaped city of a trillion suns.

universe: the chilling realization that Earth is but a mote of dust adrift in the ocean of space. The fact that Earth harbors creatures who are able to contemplate their place in the cosmic scheme must make our dust speck at least a little special. But wondering who else is out there only deepens the almost mystical enchantment of those remote celestial orbs.

Naturalists of the Night

Since I first became fascinated with the cosmos more than 40 years ago, humankind's knowledge of the universe has expanded enormously. It has been almost a continuous intoxicant for me as the discoveries have come thick and fast: quasars, pulsars, black holes, volcanic moons, bodies made entirely of ice orbiting larger worlds of liquid hydrogen. Hardly a month goes by without something new to ponder.

The mental exercise of grappling with the vast distances and sizes of celestial objects is captivating in itself. But to be able to stand under those stars and planets on a dark night is what makes backyard astronomy an addictive pastime. For me, it is communing with nature on a grand scale. I have come to know those remote stars and galaxies. The stellar panorama comes alive when one can recall: "There's the star that is 250 times bigger than our Sun...over there, in a spot I can cover with my fingernail, is a cluster of 500 galaxies, each like the Milky Way ...and over here is the nucleus of our galaxy, just behind that rift in the Milky Way." All this can be seen and appreciated with the unaided eye. One becomes a naturalist of the night, an expert in the subtleties that distinguish one celestial object from another. The experience is both humbling and exhilarating.

For thousands of astronomy enthusiasts, the ultimate in self-discovery is the exploration of some of these celestial wonders with a backyard telescope. My telescope has shown me the galactic nurseries where stars are born and the gaseous tombstones they leave behind when they die. One night not long ago, my telescope, which is typical of those used by amateur astronomers, revealed the delicate spindle-shaped image of a galaxy 70 million light-years away. The light I saw from that remote continent of stars had been hurtling through space at one billion kilometers per hour since the time dinosaurs ruled the Earth. Backyard astronomy is just as much a cerebral as a visual adventure.

When I was a teenager in the 1950s, I looked longingly at the telescope ads in *Sky & Telescope* magazine. But it was just a dream. Few astronomy buffs of that era could afford commercially produced telescopes, which were generally handcrafted and very expensive, so most decided to make their own. Some of these homegrown instruments were of exceptional quality, although many—my own included—were not much more than glorified junk. Today, with recreational astronomy booming, compared with its underground status in the '50s, the situation is completely different. Telescope making is now

Left: When Comet Hale-Bopp was at its brightest in early April 1997, it was clearly visible to the naked eye as it displayed its twin tails to millions of onlookers on planet Earth. Above: The Moon was almost completely immersed in the coppery veil of the Earth's shadow during the lunar eclipse of March 23, 1997. Backyard astronomers are ready and waiting when easily visible celestial events like these occur.

a minor component of the hobby. Dozens of different models of good-quality telescopes are available for less than $1,000, while $5,000 will buy a telescope that can outperform virtually anything used by astronomy hobbyists a generation ago.

When viewing the Moon through a top-quality 6- or 8-inch telescope (the sizes often used by amateur astronomers today), one can detect features the width of a large football stadium and subtle ripples only tens of meters in height on the lunar plains—views corresponding to those seen out the window of a spacecraft only a few hundred kilometers above the Moon's surface.

When you turn a typical amateur telescope to Jupiter, the mighty planet's ocean of clouds exhibits belts and zones in yellow, salmon, gray, white and brown. Jupiter's four large moons circle the giant like obedient servants dithering to please the master. Brilliant Venus, so dazzling to the eye, offers phases like the Moon's. Then there is Saturn's exquisite system of rings casting plainly evident shadows on the great planet.

The Starry Realm

Beyond the solar system, a limitless hunting ground awaits. A telescope or even a pair of binoculars transforms the gauzy Milky Way into a glittering array of individual stars. Elsewhere, the differing colors of the stars are on display, from pulsating variables that occasionally shine with a blood-red hue to double-star systems, where one sun may be orange, the other blue.

Star clusters ranging from loose collections of 10 or 20 suns to colossal swarms of hundreds of thousands of stars are out there, if you know where to look. And if you probe even deeper into space, thousands of galaxies—colossal star cities far beyond our own Milky Way system—dot the void like puffs of frosty breath frozen in time.

Although this celestial showcase has always been there for anyone who has access to a good telescope, only recently have such instruments been in the hands of large numbers of amateur astronomers. This book is designed primarily for those with a budding interest in the cosmos who have yet to purchase a telescope. By untangling the jargon and avoiding what I have found to be unnecessary technical baggage, this book will enable the uninitiated to make intelligent decisions about pursuing backyard astronomy and purchasing a telescope that will provide a lifetime of cosmic excursions.

But the book does not overlook the surprising variety of celestial phenomena revealed to the unaided eye: the diaphanous auroral curtains that sometimes drape the northern sky, the waltz of two planets as they slip past one another in their paths around the Sun, another galaxy of stars like the Milky Way two million light-years away, and more. Exploring the night sky is, in many ways, like a sight-seeing tour of exotic foreign lands. But, as with any journey, the appreciation of that tour is greatly enhanced when the tourist has prepared for the venture. Once experienced under the right conditions, the cosmic panorama will tempt again and again.

In December 1997, all the planets happened to be strung in a line to the left of the Sun, although only four were obvious to the unaided eye. Three of them are shown in this scene—Venus, the brightest, Mars to its lower right and Jupiter at upper left.

CHAPTER TWO

THE UNIVERSE IN ELEVEN STEPS

The universe begins to look more like a great thought than a great machine.

Sir James Jeans

Astronomy stretches the mind like nothing else in human experience. Enormous distances, sizes and timespans challenge the imagination; alien concepts like black holes and galactic cannibalism strain comprehension. Yet it is possible to bring the total picture—the structure and extent of the entire universe—into focus.

When our great-grandparents were children, no one knew what, if anything, existed beyond the visible stars. The discovery of the universe beyond was astronomy's 20th-century enterprise. Astronomers now have reason to believe that there is an end to space and time, that the universe is finite. Ideally, I would like to be able to display the various components and the scale of the entire universe in a single illustration, but that would require a whole wall. The only way to achieve a realistic perspective in book form is through a series of ever-widening views.

The illustrations in this chapter, rendered by astronomical artist Adolf Schaller, are scientifically reasonable representations of the universe as it is currently understood. Each of the drawings represents a cubic volume of space a million times larger than the one before. That is, in each increment, the width, height and depth are expanded by a factor of 100, giving a volume increase of one million. Throughout the sequence, the cube remains centered on Earth.

We begin with Earth (facing page) and enclose it within an imaginary cube whose sides are slightly larger than the diameter of the planet. This is a manageable starting point. The Earth's dimensions can be easily grasped. Many people routinely fly one-quarter of the way around the globe in a few hours. In a high-flying jet liner, it is possible to get a hint of the Earth's surface curvature at the horizon, while overhead, the deep blue color signals the thinning atmosphere at the edge of space.

The next step outward reduces Earth to

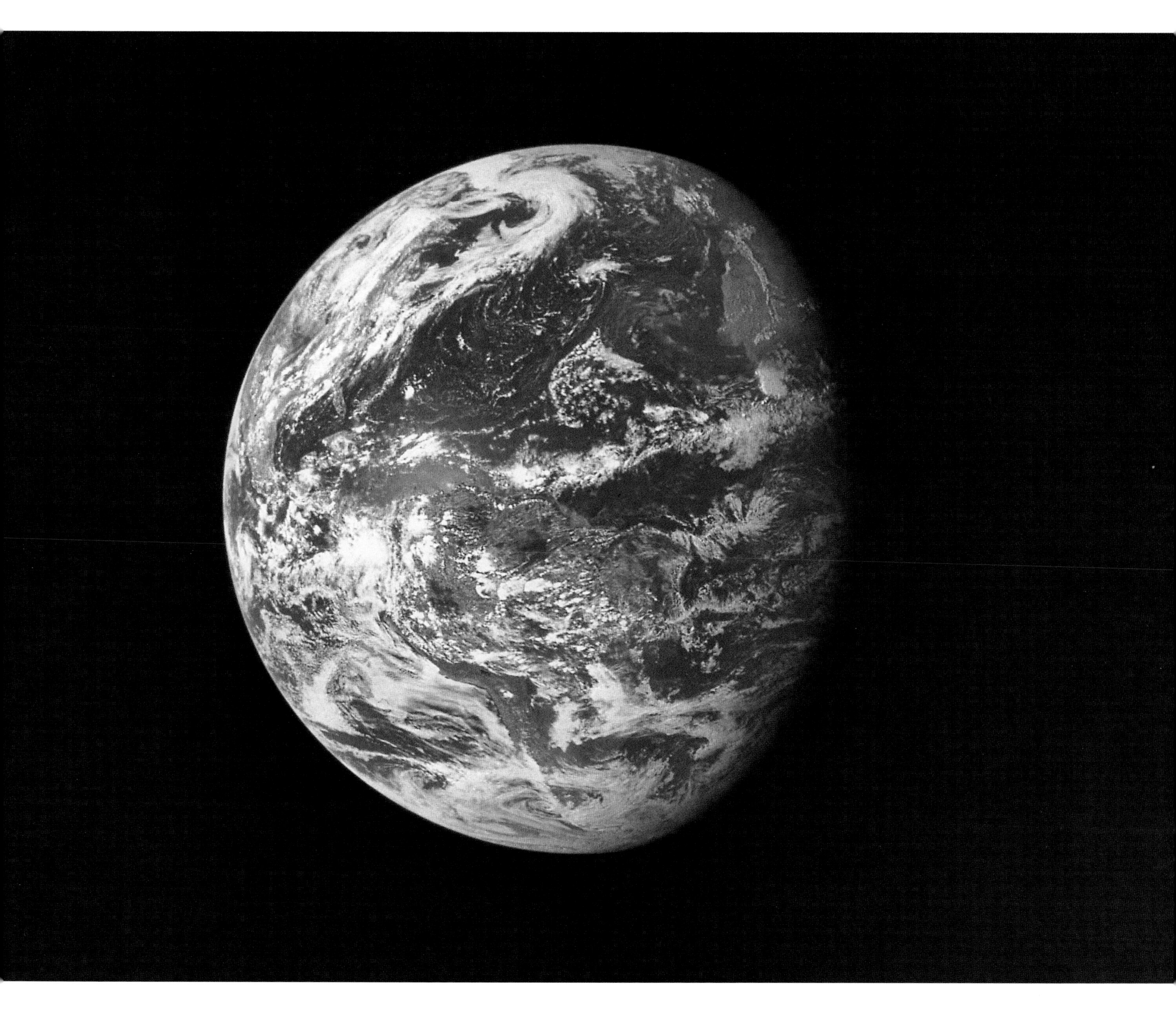

Step 1 The first of the 11 steps places Earth in an imaginary cube only slightly larger than the planet itself. Earth is 12,742 km in diameter; its mass equals 6 billion trillion tons. Yet it is but a mote of dust in the universe, as the widening perspectives in the following steps reveal.

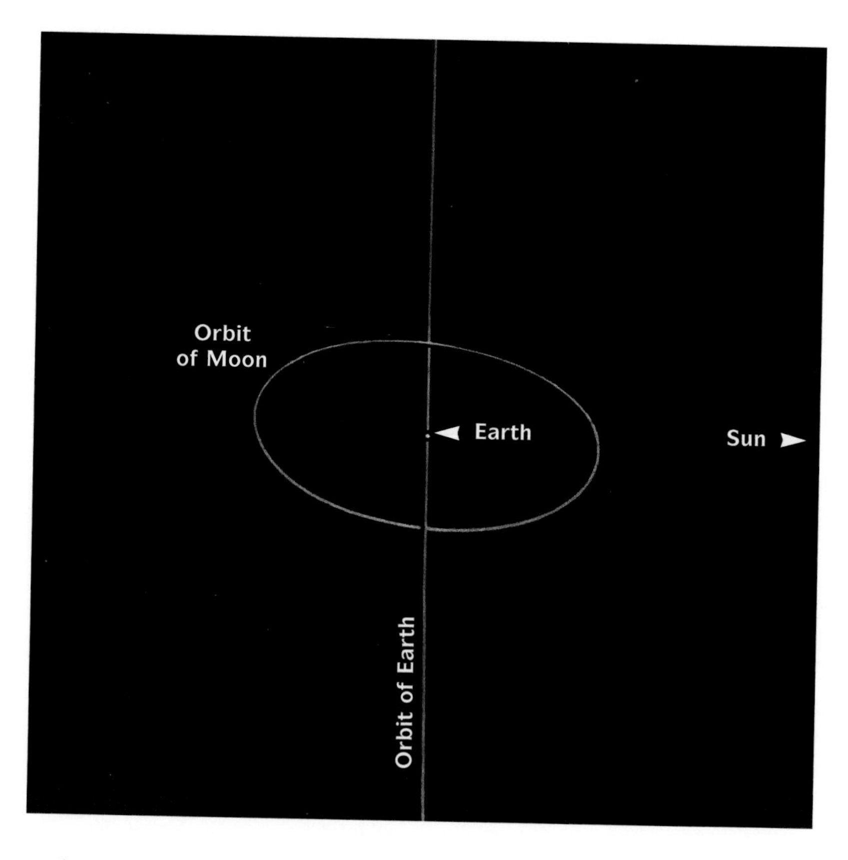

a dot, since each side of the cube has expanded 100 times, to about two million kilometers. The cube now comfortably contains the Moon's orbit, an 800,000-kilometer-diameter circuit around Earth. The distance from Earth to the Moon is almost exactly 30 Earth diameters, a modest leap that required two days for the Apollo astronauts to span. The fastest interplanetary spacecraft can traverse it in about six hours. No other celestial object, except the rare errant boulder from the asteroid belt, ever comes within the bounds of this second cube.

The third cube, centered on Earth, includes portions of the orbits of Mercury, Venus and Mars, our nearest planetary neighbors. Dozens of robot spacecraft have navigated the gulf between these worlds. This cube, roughly 200 million kilometers across, is a little larger than a standard cosmic measuring unit known as an astronomical unit, or AU. One AU, 150 million kilometers, is the Earth-Sun distance.

The region between these planets is essentially empty space, except for a few thousand renegade asteroids—chunks of rocky material left over from the formation of the planets—that have drifted in from the asteroid belt beyond Mars. The largest of these are flying mountains that pose the threat of mass destruction if one ever hits Earth. Fortunately, such collisions are exceedingly rare. The last big impact occurred 65 million years ago, around the time of the extinction of the dinosaurs.

Other intruders in our basically quiet neighborhood are comets, also mountain-sized but largely composed of water ice. When their surfaces are vaporized by solar radiation, comets sprout the familiar filmy tails seen in photographs. Apart from these visitors from other parts of the solar system, our sector of

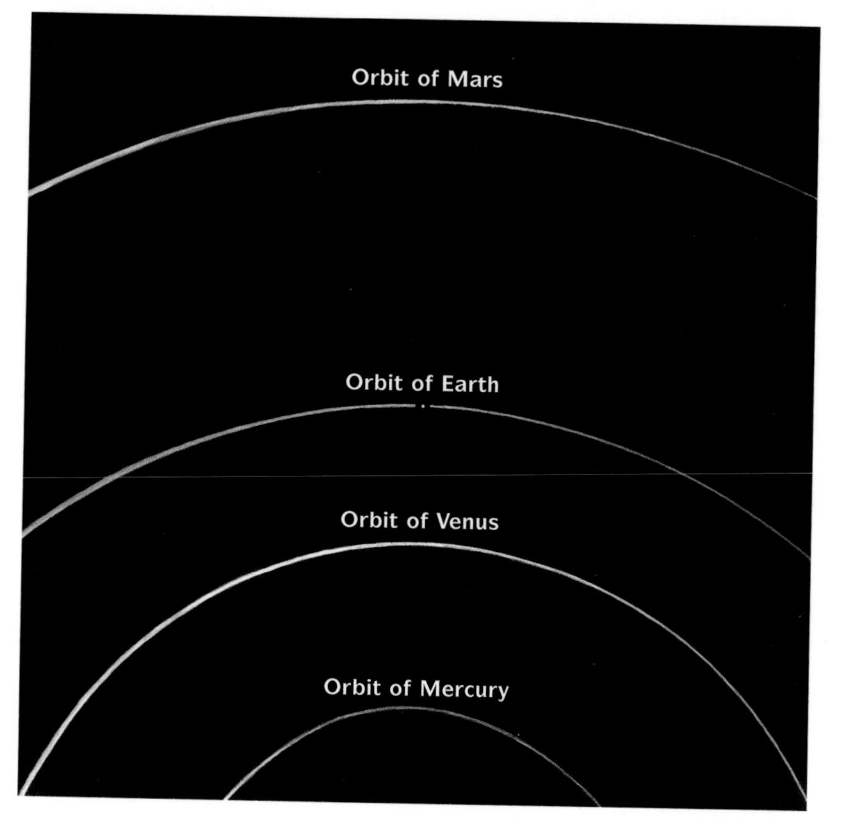

Step 2

Width of cube: 2.0 million km
Light time across cube: 7 seconds
Volume of cube: 8 million trillion cubic km
Distance from Earth to Moon: 384,500 km (average)
Orbital period of Moon: 27.32 days
Diameter of Moon: 3,476 km
Mass of Moon: 1.2% of Earth   Volume of Moon: 1.6% of Earth

Step 3

Width of cube: 1.2 AU, or 200 million km
Light time across cube: 13 minutes   Volume of cube: 1.8 cubic AU
Minimum distance from Earth to Venus's orbit: 0.27 AU
Minimum distance from Earth to Mars's orbit: 0.38 AU
Minimum distance from Earth to Mercury's orbit: 0.53 AU
Orbital period of Mercury: 88.0 days   Orbital period of Venus: 224.7 days
Orbital period of Earth: 365.25 days   Orbital period of Mars: 687.0 days

space is relatively tranquil and uneventful. It must have been this way for some time: Earth has orbited the Sun nearly five billion times since it formed and is still here.

The Earth's companion planets, endlessly swinging in their gravitationally prescribed paths about the Sun, are a varied lot, ranging from crater-pocked Moonlike Mercury to colossal Jupiter, a globe of liquid hydrogen whose bulk equals 1,000 Earths. If there is one overriding lesson that has been learned from the interplanetary space probes, it is that these other worlds are more alien than have ever been imagined by the most vividly creative science fiction writer.

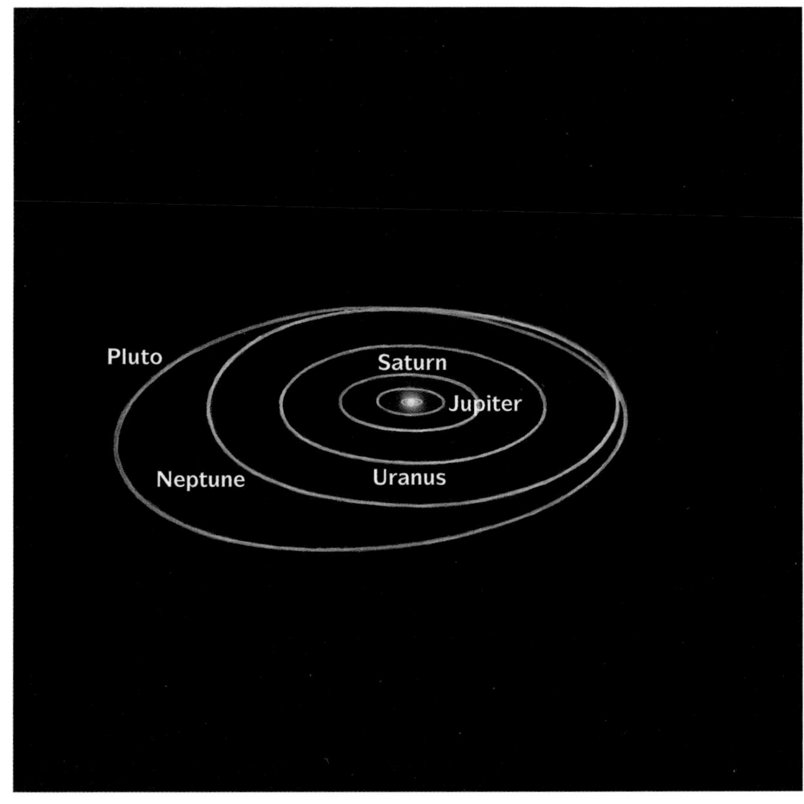

To include all nine planets in the solar system out to Pluto, we take the fourth increment in our progression—a cube 120 AU wide. The only solar system members left outside this cube are comets, whose elongated orbits usually keep them billions of kilometers from the Sun. (Pluto's orbit passes inside Neptune's, but the intersection points are like freeway interchanges, with one orbit well above the other.)

As seen from the edge of this fourth cube, all the planets except Jupiter and Saturn would be invisible to the unaided eye. Even those two would be inconspicuous specks. On this scale, the blazing Sun is the only object of significance; the planets, by comparison, are merely debris orbiting around it.

When Earth-based astronomers look at other stars, even using the most powerful telescopes, no planets can be seen directly. Like the Sun, the stars overpower their planetary systems,

Width of cube: 120 AU Light time across cube: 17 hours **Step 4**
Volume of cube: 1.7 million cubic AU Orbital period of Jupiter: 11.86 years
Orbital period of Saturn: 29.46 years Orbital period of Uranus: 84.0 years
Orbital period of Neptune: 164.8 years Orbital period of Pluto: 248.0 years
Distance from Sun to: Mercury, 0.39 AU; Venus, 0.72; Earth, 1.00; Mars, 1.52; Jupiter, 5.20; Saturn, 9.54; Uranus, 19.2; Neptune, 30.1; Pluto, 24.6 to 52.6*
*Pluto's elliptical orbit places it closer to Sun than Neptune is from 1979 to 1999

Width of cube: 12,000 AU, or 0.19 light-year **Step 5**
Light time across cube: 70 days
Volume of cube: 1.7 trillion cubic AU, or 0.007 cubic light-year
Estimated number of comets inside (or just beyond) cube: 100 billion, ranging from 1,000 km to just a few meters in diameter
Estimated total mass of all known comets: 100 times mass of Earth

masking even worlds as great as Jupiter. However, during the 1990s, astronomers indirectly detected Jupiter-sized planets orbiting around more than a dozen Sunlike stars. Telltale shifts in the spectra of these stars—shifts caused by the orbiting planet's gravity pulling the star alternately toward and away from Earth—have revealed the existence of other solar systems.

The fifth step outward produces a 12,000-AU cube, enclosing almost exclusively empty space. Pluto's orbit has shrunk to a tiny oval near the central Sun. From the edge of the cube, the Sun would appear as simply a very bright star. A haze of comets known as the Oort cloud—named after Jan Oort, the Dutch astronomer who first suggested that billions of comets roam the fringes of the solar system—is emphasized for clarity in the illustration. As large as this cube is, the nearest star is still 50 times farther away.

From now on, miles or kilometers become useless for scaling distances. Even astronomical units will soon be cumbersome. For tabulating interstellar distances, astronomers use light-years, the distance that light travels in a year at its constant velocity of 7.2 AU per hour (299,792 kilometers per second).

Our sixth increment takes us to a cube 1.2 million AU wide, or, more conveniently, 20 light-years. The Sun now takes its place as one star among many. Our nearest neighbor, Alpha Centauri, is a triple-star system 8,000 times farther from Earth than frigid Pluto. Distances between star systems are awesome. If the Sun were the size of a grapefruit, this cube would have to be larger than Earth to provide the correct scale.

A comparison of our neighbor stars reveals that the Sun is much brighter than the average star. The majority of stars

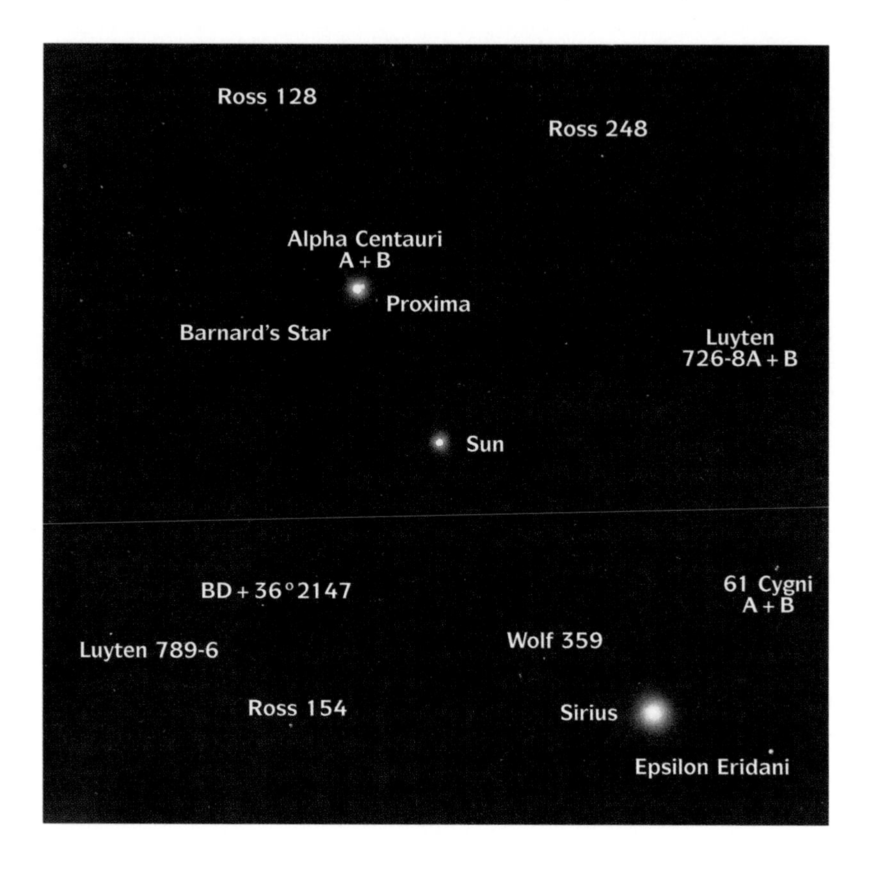

Step 6

Width of cube: 1.2 million AU, or 20 light-years
Volume of cube: 8,000 cubic light-years
Number of stars in cube: 17 (8 single stars; 3 double systems; 1 triple)
Actual brightness of some nearby stars compared with Sun: Alpha Centauri A, 1.2; Alpha Centauri B, 0.36; Proxima Centauri, 0.00006; Barnard's Star, 0.00044; Sirius, 23; Epsilon Eridani, 0.3; 61 Cygni A, 0.08; 61 Cygni B, 0.04

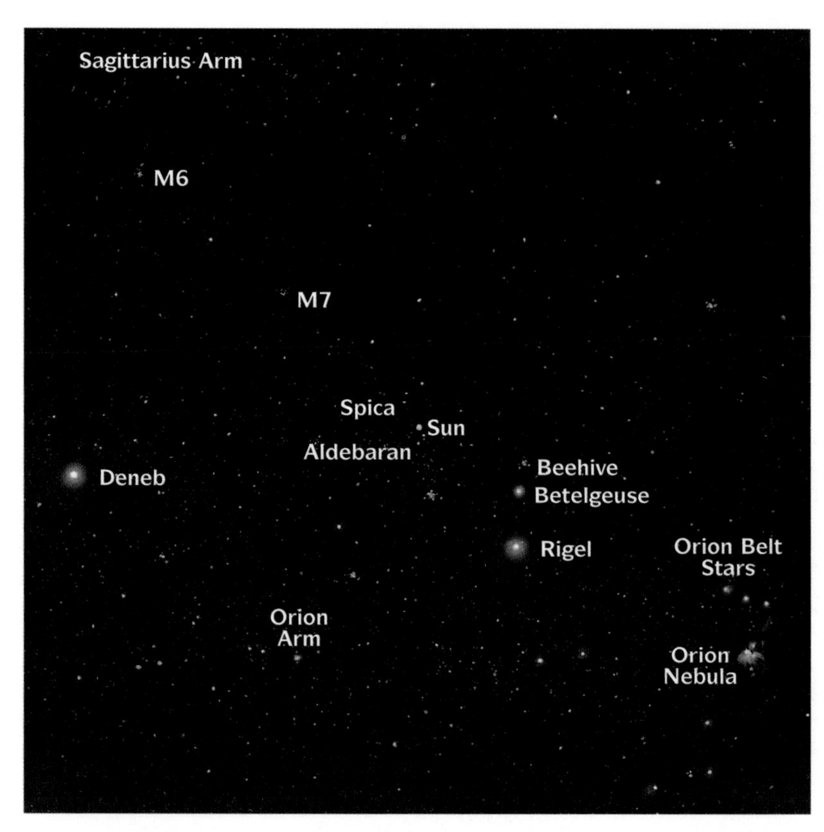

Step 7

Width of cube: 2,000 light-years
Volume of cube: 8 billion cubic light-years
Estimated number of stars in cube: 2 million
Distance from Sun to: Aldebaran, 65 light-years; Spica, 260; Pleiades, 370; Betelgeuse, 430; M7 cluster, 800; Orion Nebula, 1,400

are dim suns astronomers call red dwarfs. Just 5 percent of all stars are brighter than our Sun. The only star in this cube that is significantly more luminous than the Sun is Sirius, the brightest star in the night sky and the nearest naked-eye star visible from midnorthern latitudes. Because of its location in the deep southern sky, Alpha Centauri is seen only from Miami or farther south.

Each star in the Sun's vicinity has about 400 cubic light-years of empty space around it, plenty of room for Captain Picard and the crew of the starship *Enterprise* to scoot around without encountering anything. Or would they? Apart from the occasional vagabond comet at the outskirts of each star's Oort cloud, is there anything there?

The space between stars is not a total vacuum, but it is close. There is less than one atom for every cubic centimeter, compared with 10 million trillion atoms in a cubic centimeter of air at sea level on Earth. The only substantial objects that might exist undetected between the stars would be Jupiter-class planets floating alone or black holes—gravity whirlpools created when certain types of massive stars explode. However, there is no evidence, either theoretical or observational, that significant numbers of either class of object are lurking between the stars.

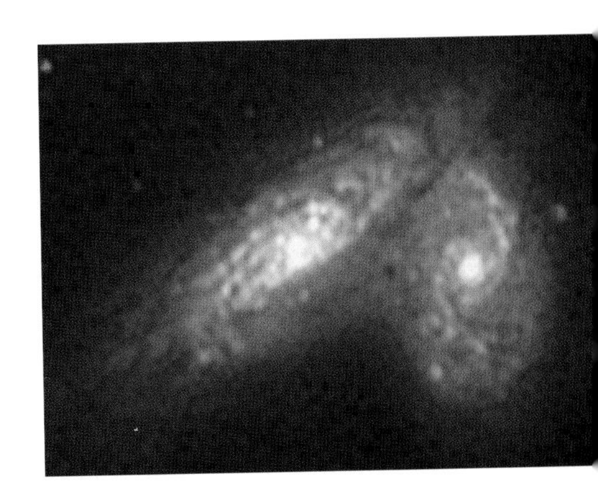

The seventh cube in our outward march, 2,000 light-years across, encloses about two million stars, several hundred times more than are visible from Earth with the unaided eye on the clearest nights. The stars seem to crowd on top of one another, but this is simply because we are showing so much in such a small space. Remember that they are all several hundred thousand AU from each other. A few prominent naked-eye

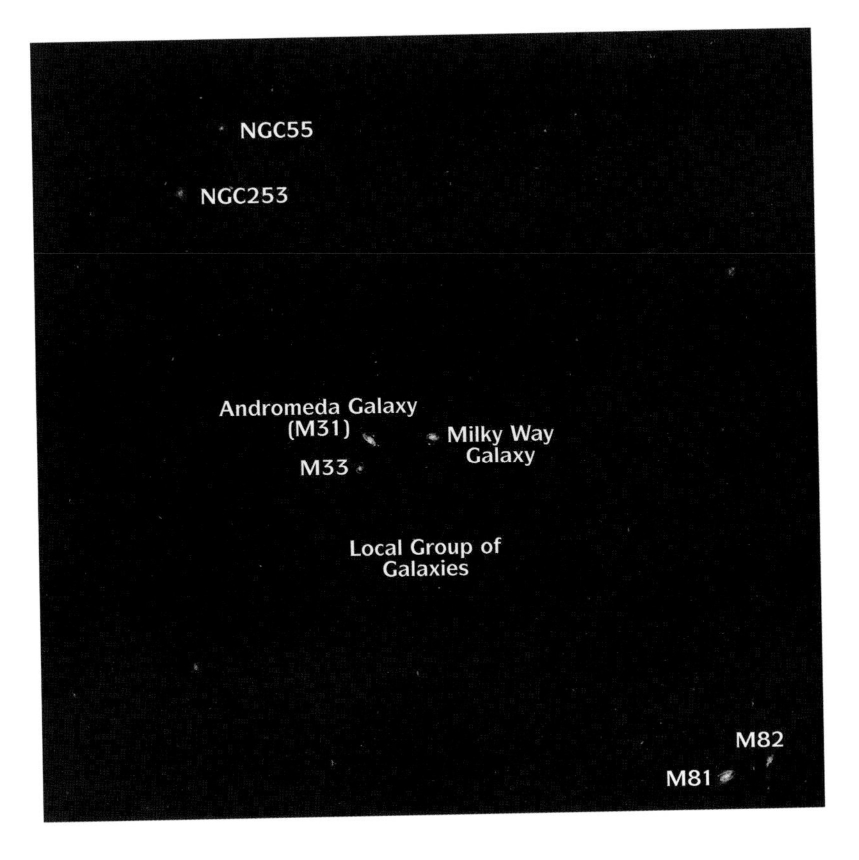

Step 8

Width of cube: 200,000 light-years
Volume of cube: 8,000 trillion cubic light-years
Estimated number of stars in cube: 1 trillion
Diameter of Milky Way Galaxy: 90,000 light-years
Orbital period of Sun around galaxy: 220 million years
Mass of Large Magellanic Cloud: 10 billion solar masses
Mass of Small Magellanic Cloud: 2 billion solar masses

Step 9

Width of cube: 20 million light-years
Volume of cube: 8 billion trillion light-years
Number of major spiral galaxies in cube: 6
Estimated number of minor spirals and dwarf galaxies: 100
Estimated number of stars in cube: 10 trillion
Cube as a percentage of entire universe: roughly 0.00000003%

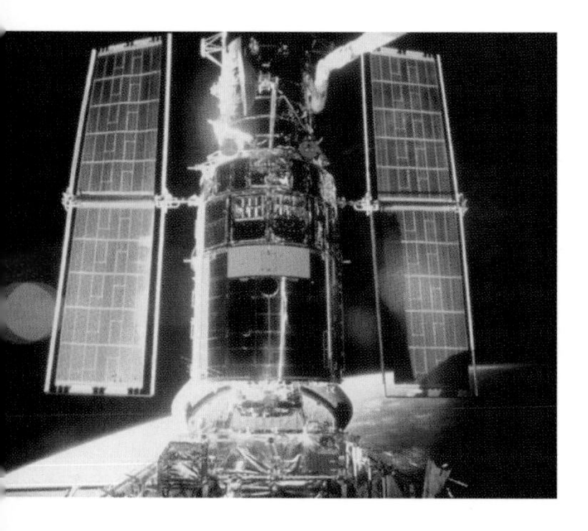

objects are identified in this seventh step, including some bright stars and the nebula in the constellation Orion as well as the Pleiades star cluster in Taurus. All are several hundred light-years distant. Although the vast majority of stars seen on a typical clear night are contained in this celestial cube, it represents only about one-thousandth of 1 percent of our star city, the Milky Way Galaxy.

The eighth cube, 200,000 light-years wide, easily spans the Milky Way Galaxy, which is 90,000 light-years in diameter. In the Sun's vicinity, roughly two-thirds of the distance from the hub to the edge, the galaxy is about 3,000 light-years thick. For the proper thickness-to-width ratio, imagine two CDs held together. Our Sun would be a speck of dust between the disks. Moving at a velocity of one AU per week, the Sun takes about 220 million years to complete one revolution around the galactic nucleus. Since the Sun and Earth were born, they have made fewer than 25 such trips.

The ninth increment in our push to the edge of the universe reveals a cube 20 million light-years on each side. Our galaxy is now simply one of several dozen speckling an enormous volume of space. The galaxies within three million light-years of the Milky Way are gravitationally bound into a permanent family that astronomers call the Local Group. Only one of these, the Andromeda Galaxy, is comparable in size to our own. The rest are less than one-twentieth as massive. Beyond the Local Group, near the limits of the cube, other galaxies similar to the Milky Way hint at what lies in the depths of space.

The space between galaxies is as close to a complete vacuum as can be imagined. Only one atom per cubic meter

Step 10

Width of cube: 2 billion light-years
Estimated number of galaxies in cube: 100 million
Estimated number of stars in cube: a few million trillion
Average distance between galaxy superclusters: 300 million light-years
Largest known galaxy superclusters: 100,000 galaxies

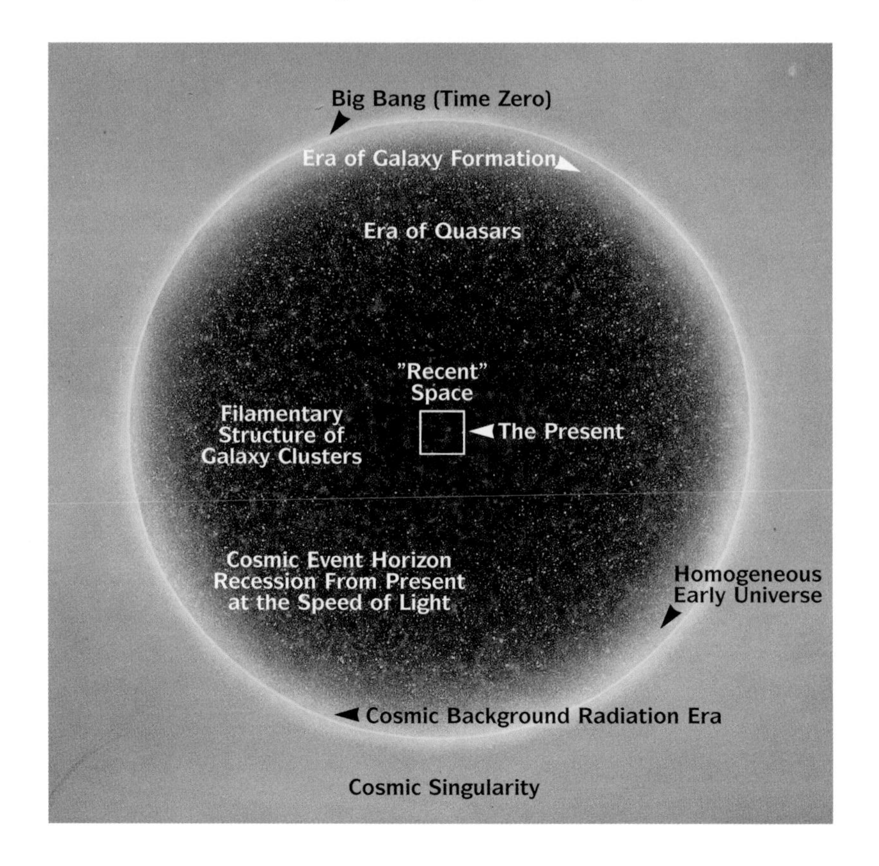

Step 11

Width of cube: greater than the known universe
Estimated number of galaxies in universe: 100 billion
Estimated number of stars in universe: 5 billion trillion
Estimated number of planets in universe: very uncertain, perhaps trillions
Estimated age of universe: 14 to 16 billion years
Evolutionary destiny of universe: continued expansion, perhaps forever

Known as the Hubble Deep Field, this 100-hour exposure by the great orbiting telescope reveals thousands of previously unknown galaxies in a sector of the sky about the size of a grain of salt held at arm's length.

The most distant galaxies seen here —barely visible as small smudges— are 5 to 10 billion light-years away. Light now reaching us from these remote galaxies began its journey before Earth even existed.

rides the intergalactic void. Long before Earth existed, matter in the universe clumped into colossal islands that eventually became galaxies. They range from monsters 100 times more massive and several times larger than the Milky Way Galaxy to midgets containing just a few thousand stars.

Galaxies are nature's greatest building blocks. Observations by the Hubble Space Telescope suggest that the universe contains at least 50 billion of them. To see how galaxies form the fabric of the universe, we move on to the tenth step, a cube of space two billion light-years across. Within this volume, millions of galaxies swim in a seemingly endless abyss. The Milky Way is lost among the galactic throng. But there is structure. Visible at last is the grand architecture of the cosmos. The galaxies are arrayed not at random but in knots, clumps and ribbons. These superclusters, as they are called, are collections of smaller galaxy clusters like the Local Group. Some superclusters contain tens of thousands of galaxies.

The supercluster to which we belong, the Virgo galaxy supercluster, has at least 5,000 galaxy members and is about 100 million light-years across. The Local Group is near one outer edge. The galaxies M81 and M82, in the lower right of the previous cube, are also members, somewhat nearer the supercluster's central zone.

The reason galaxies are congregated in tattered but discrete clusters, rather than being haphazardly dispersed, is still uncertain. One theory suggests that soon after the universe formed, its matter swirled into ribbon- and pancake-shaped slabs which later became the birthplaces of the galaxies.

There is some fairly persuasive evidence that our universe does not go on forever. Current estimates suggest that the edge is about 14 billion light-years beyond the limits of this supercluster cube. We now take the audacious step of one final extension in our cosmic progression, this time by a linear factor of 20, to enclose the entire known universe.

The suggestion that the universe really ends somewhere is based on the idea that the universe has existed for a finite period. According to the widely accepted Big Bang theory, the universe came into existence about 15 billion years ago in a colossal genesis explosion. It has been expanding at near light speed ever since. In the time it takes to read this sentence, the universe will increase in volume by 100 trillion cubic light-years.

The expansion separates galaxy clusters from one another, like dots on an inflating balloon. This process has made it

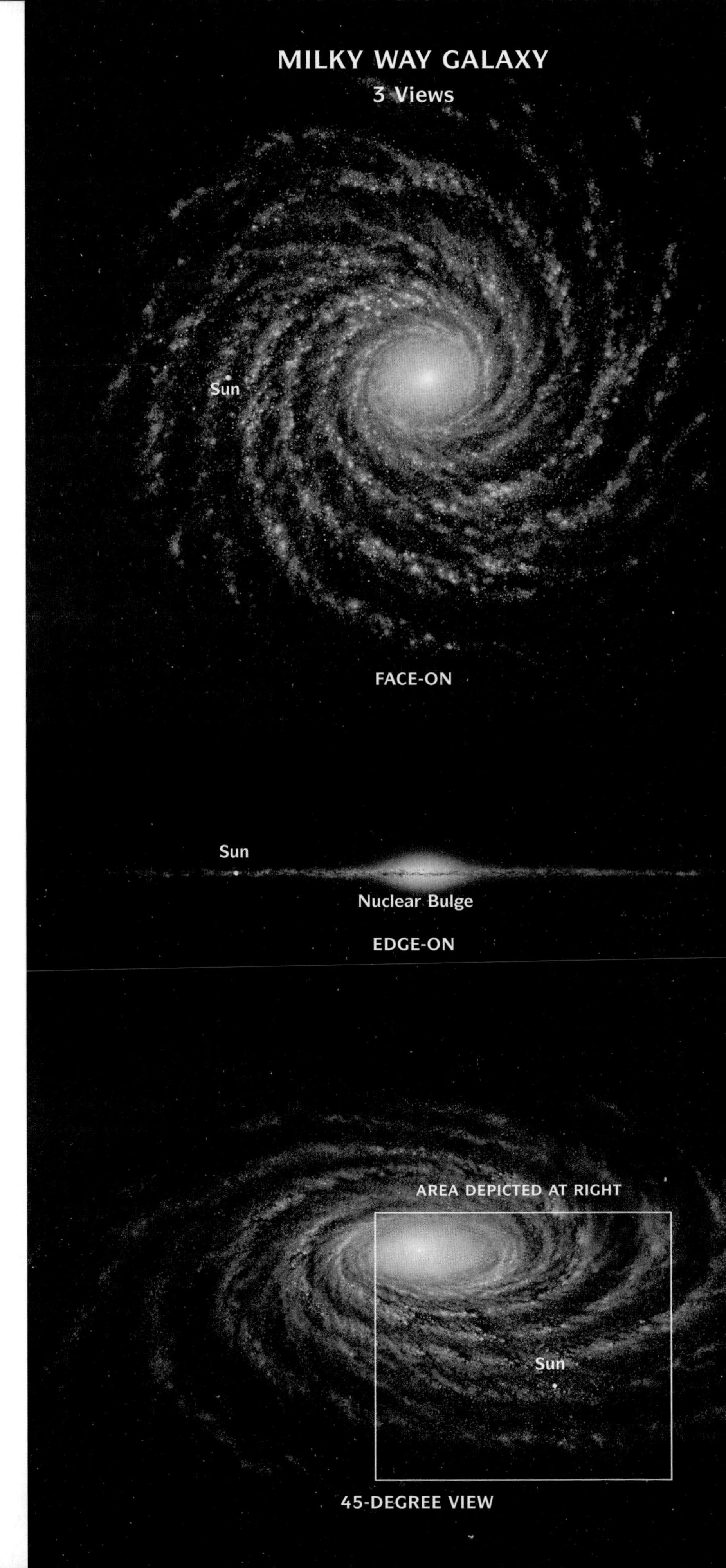

Three views of our galaxy display the shape and structure of the vast 90,000-light-year-wide system. Close-up of the Sun's quadrant of the galaxy, facing page, pinpoints our location and that of a few well-known stars and nebulas.

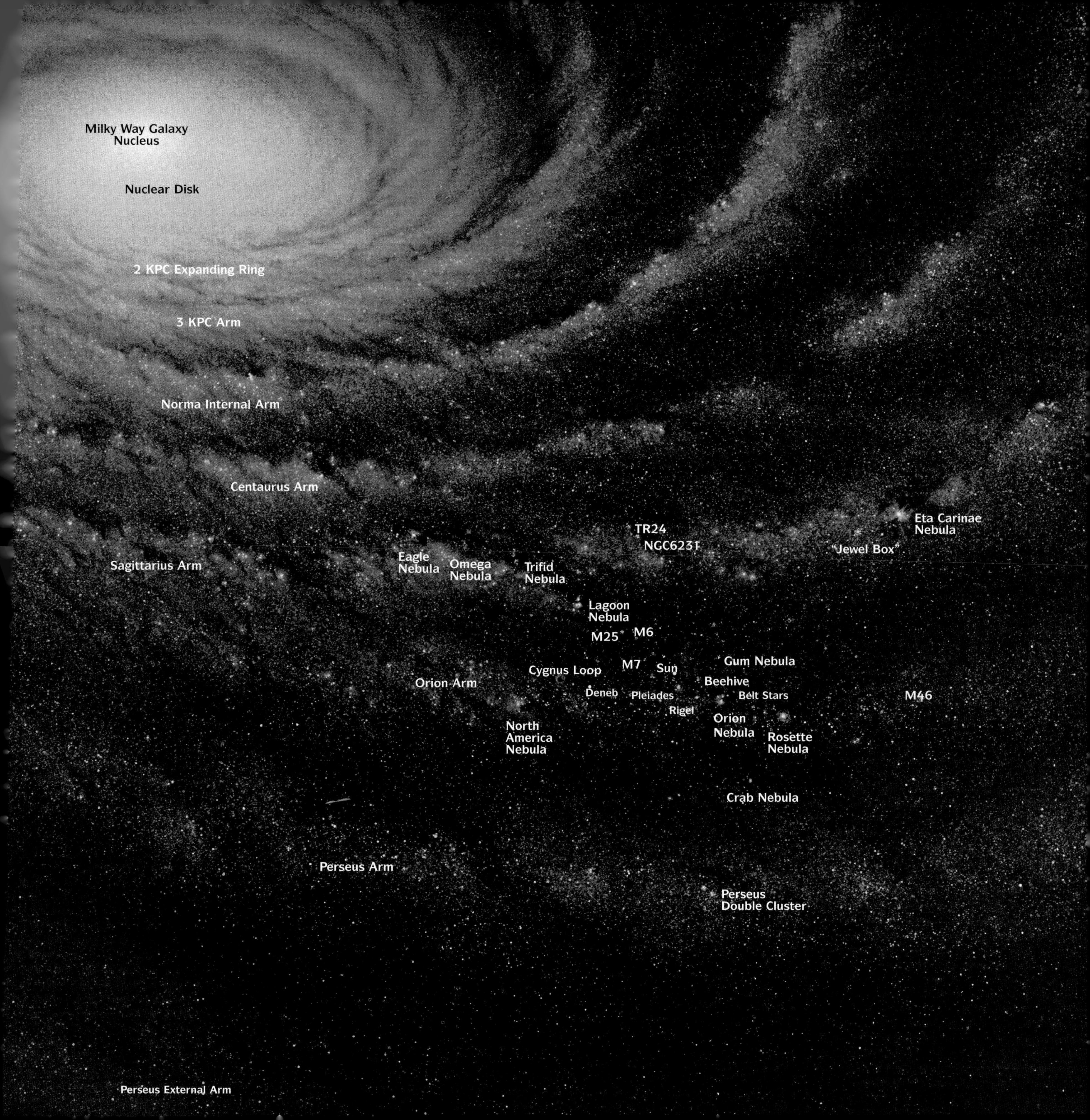

Milky Way Galaxy
Nucleus

Nuclear Disk

2 KPC Expanding Ring

3 KPC Arm

Norma Internal Arm

Centaurus Arm

Sagittarius Arm

Eagle
Nebula

Omega
Nebula

Trifid
Nebula

TR24

NGC6231

Eta Carinae
Nebula

"Jewel Box"

Lagoon
Nebula

M25

M6

M7

Sun

Gum Nebula

Cygnus Loop

Orion Arm

Deneb

Pleiades

Rigel

Beehive

Belt Stars

M46

North
America
Nebula

Orion
Nebula

Rosette
Nebula

Crab Nebula

Perseus Arm

Perseus
Double Cluster

Perseus External Arm

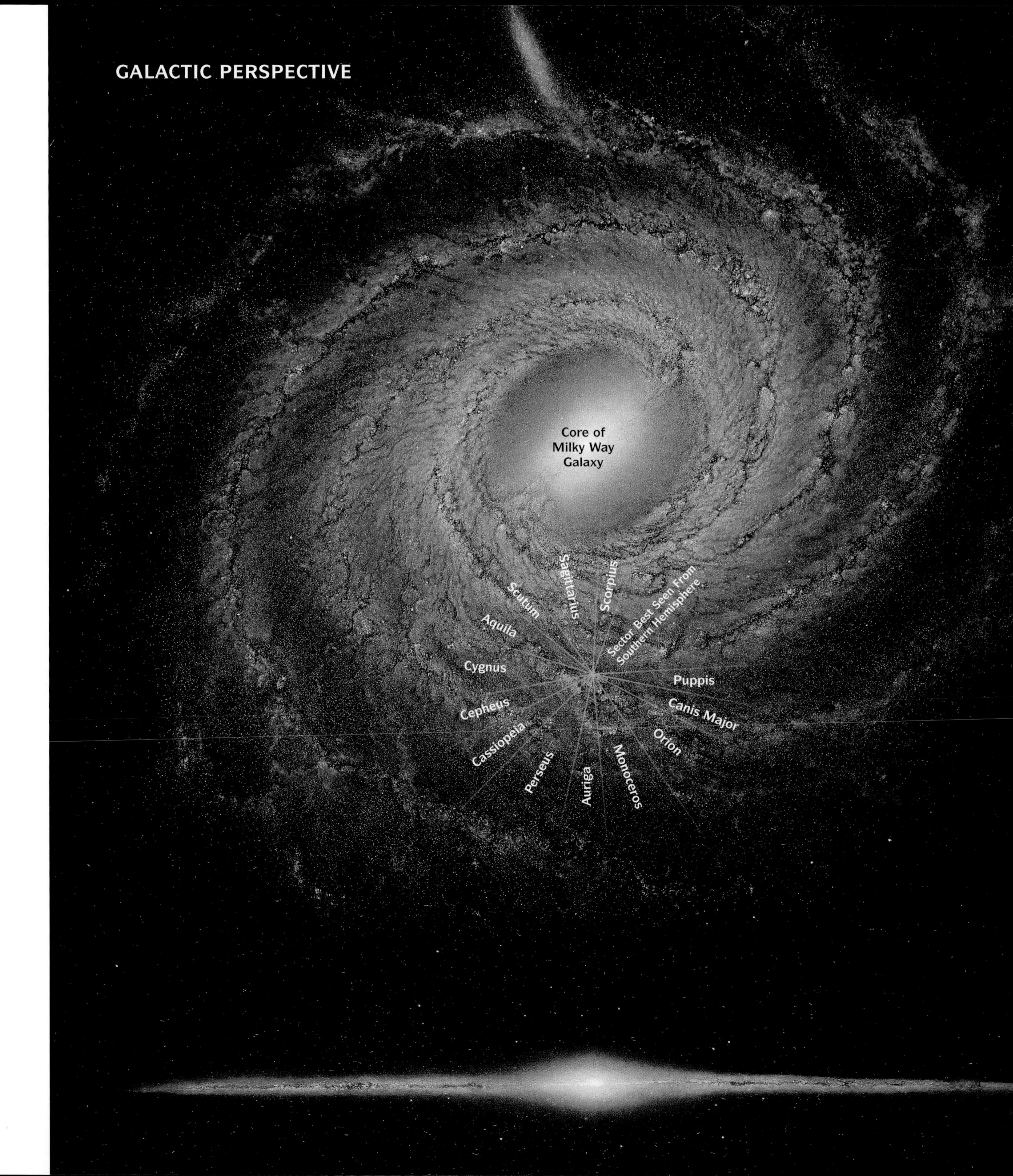
GALACTIC PERSPECTIVE
Core of Milky Way Galaxy
Sagittarius
Scorpius
Scutum
Sector Best Seen From Southern Hemisphere
Aquila
Cygnus
Puppis
Cepheus
Canis Major
Cassiopeia
Orion
Perseus
Auriga
Monoceros

possible for astronomers to probe into the past, all the way back to the creation era. The expansion has pushed us so far from some sectors of the universe that it takes light from remote galaxies vast amounts of time to reach Earth. These galaxies are seen as they were when the light left, not as they are today. Gazing huge distances into space is therefore like traveling in a cosmic time machine, an enormously useful tool with which to study the evolution of the universe.

The Milky Way Galaxy is thought to be about 13 billion years old. When we look at objects that are 10 to 13 billion light-years away, we see them as they were soon after their formation (assuming that all galaxies were born at roughly the same time). Many of these youthful galaxies are far more energetic than our sedate star city, pumping out up to 10,000 times the radiation emitted by the Milky Way Galaxy. No matter where astronomers look at these vast distances, vigorously energetic objects are seen. Astronomers assume that they are witnessing the earliest—and, apparently, the most violent—stages of galaxy formation a few billion years after the creation of the universe.

When instruments probe to 15 billion light-years, they record nothing but a dull energy haze, called the microwave background. This is believed to be the remnant "glow" of the creation explosion. Thus astronomers conclude that we inhabit a universe whose observational extent is defined by its age. The edge of the universe is an edge in time; it is not possible to look back past the beginning.

Exploration of these distant realms billions of light-years away is beyond the capability of amateur equipment. Only the giant instruments of the largest research observatories can be used to seek out the secrets at the edge of space and time. Remarkably, it is just this final "total universe" cube and the supercluster cube before it that are largely off-limits to backyard skywatchers. Even with the unaided eye, it is possible to see the Andromeda Galaxy, more than two million light-years into Cube 9. A small telescope can reveal the brightest galaxies throughout Cube 9 and even a few that are hundreds of millions of light-years into the inner part of Cube 10, but the cosmic hunter must know where to look.

The illustration at left, showing a face-on and an edge-on view of the Milky Way Galaxy, is based on our best current estimate of what our galaxy looks like. Because we are located inside the galaxy, its exact appearance has been inferred from analyses of optical, infrared and radio-telescope surveys. The elongated feature near the top of the face-on view is a recently discovered small galaxy being absorbed into the main body of the Milky Way. The wheel of wedges centered on the Sun identifies the constellations we see as we look out into our own galaxy. Refer back to this illustration when you become familiar with them. The three photographs on this page are galaxies within a distance of 60 million light-years that resemble our galaxy. Alien astronomers observing from a planet of a star in a remote galaxy would see the Milky Way pretty much as we see these galaxies.

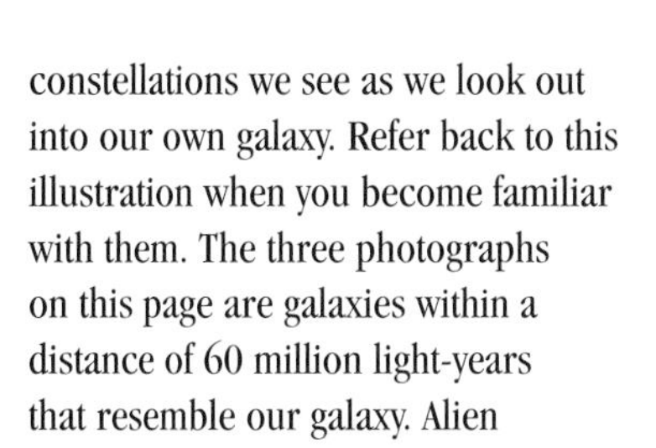

The Milky Way Galaxy

The Milky Way Galaxy is our hometown in the universe. By cosmic standards, it is a major metropolis, with several hundred billion stellar citizens—there are at least 40 times as many stars in our galaxy as there are humans on Earth.

The disk-shaped galaxy, about 90,000 light-years in diameter, has a brilliant nuclear bulge, roughly 10,000 light-years thick and 13,000 light-years wide, containing at least 100 billion stars. In the nuclear region, 10 to 1,000 stars occupy the same volume of space that our Sun does in one of the spiral arms. Stars close to the nucleus are less than one-quarter of a light-year apart; near collisions must be common, whereas the Sun has probably never come within one-quarter of a light-year of another star in at least a billion years.

The galaxy's system of spiral arms emerges from the nuclear bulge into a flat, symmetrical pinwheel. The Sun is located about two-thirds of the way out from the galaxy's center, on the inside edge of one of the spiral arms, known as the Orion Arm. This is the galaxy's suburban area, well away from the denser nucleus but not too far away

We see our home galaxy, the Milky Way, from the inside as a misty band arcing overhead on summer nights. This wide-angle view, taken from an Arizona mountaintop, clearly shows the galaxy's hub as a bulge above the southern horizon.

from stellar neighbors. The most easily visible of those neighbors are the blue giants and supergiants, such as the stars in Orion's belt. These enormously powerful stars, significantly more massive than the Sun, define the spiral arms and give them their bluish white hue.

The yellowish tone of the galaxy's nuclear regions is the result of a a preponderance of yellow and red stars and a lack of blue giants. Blue giants are young stars, and the bulge seems to be the realm of more mature suns. Blue giants live fast and die young; they squander their stellar fuel at prodigious rates and last for only a few million years. After a short lifetime of blazing glory, blue giants evolve to red giants and, soon after, leave the scene, possibly in the blast of a supernova.

Blue giants, along with lesser stars like the Sun, are born in nebulas —dark clouds of gas and dust woven throughout the galaxy's spiral arms. Regions where stars are actually forming, such as the Orion Nebula, dot the spiral arms as bright nebulas that appear pinkish in color photographs. Several star-forming nebulas visible in binoculars are labeled in the close-up of our quadrant of the Milky Way (page 21). That illustration also identifies prominent stars and star clusters, most of which are visible to the unaided eye.

Star clusters are the next evolutionary stage after the bright nebulas. As time passes, the clusters usually disperse, the individual stars spreading throughout the spiral arms. There are nearly as many stars *between* the spiral arms as in them. But the interarm stars are almost all less luminous types, similar to or fainter than the Sun. The spiral arms appear to be a shock front or density wave in which star formation occurs, the wave somehow propagating throughout the galactic disk. The blue giants and supergiants that define the arms do not live long enough to drift into the interarm zones.

As the Sun and other stars orbit the nucleus of the galaxy, the spiral-arm density wave sweeps over them, initiating new-star formation. Our Sun and planetary system seem to be just entering the Orion Arm. Many of the stars in our immediate vicinity are members of a pocket of fairly young stars called Gould's belt, inside the Orion Arm. When we look beyond Gould's belt to the stars in the constellation Orion, we are looking down the Orion Arm toward its trailing edge. When we sight in the direction of the constellation Perseus, we are looking away from the nucleus toward the Perseus Arm (page 22). The constellation Cygnus is located in the inward segment of the Orion Arm, in the opposite direction from the constellation Orion. In late summer, the Milky Way near the southern horizon is the Sagittarius Arm, toward the galaxy's center. Dark nebulas obscure the nuclear bulge and limit our view much beyond the Sagittarius Arm.

Beyond the spiral arms is the galactic halo, the abode of the globular clusters—dense, spherical swarms of up to four million suns. The globulars have huge looping orbits around the galaxy's nucleus. Recent research indicates that the halo—present around other galaxies as well—contains the equivalent mass of hundreds of billions of stars. But far fewer stars are seen. This "invisible" mass is one of modern astronomy's major mysteries.

The renderings of the structure of the Milky Way Galaxy in this chapter are based on the most up-to-date findings. But since a direct visual view of the nucleus is blocked by nebular material, other methods—primarily radiotelescope analyses—have filled in the details. Even so, our best picture of the overall structure of our own galaxy is not nearly as clear as routine observatory photographs of other galaxies.

In the winter sky, right, we see the dimmer, less structured part of the Milky Way Galaxy that lies opposite the central hub. Top: The galactic-core region lies between the brightest stars of the summer constellations Sagittarius and Scorpius.

CHAPTER THREE

BACKYARD ASTRONOMY

We had the sky up there,
all speckled with stars,
and we used to lay on our backs and look up at them
and discuss about whether they was made or only just happened.

Mark Twain
Huckleberry Finn

More than a century ago, Ralph Waldo Emerson wrote: "The man in the street does not know a star in the sky." No surveys have ever been conducted to determine by what percentage Emerson was right, but Shakespeare pointed out part of the problem in *Julius Caesar* when he described the night sky as "painted with unnumbered sparks."

Despite its magnificence, the starry night sky can be a bewildering chaos of luminous points. It takes time to sort it out. For every curious stargazer who has learned to distinguish one star from another, there are dozens who have taken a star chart outside and, after an hour or two, have given up in frustration.

The problem is usually the charts, rather than the observer. Although not in use much today, totally impractical charts, with mythological characters and creatures drawn over the star patterns, are still around. Even modern star charts and so-called star finders are often either too small or too crowded with dots and lines to be useful during the crucial first nights of star identification.

I vividly remember the thrill of recognizing my first constellation on a clear winter night in 1958. (Constellations are patterns of stars long ago partitioned and named by our ancestors.) I had seen constellation diagrams in astronomy books, but I was intimidated by the complex-looking charts. Then, as I gazed into that crisp winter sky, one image finally crystallized. Standing there before me was the mighty celestial hunter Orion, his glittering three-star belt framed by four prominent stars marking the shoulders and legs of the legendary nimrod. Every winter since then, Orion has been a familiar friend during the five months or so that he strides across the evening sky.

Orion is the second most prominent stellar configuration in the night sky. Most easily recognized is the Big Dipper, whose seven stars are a familiar sight to many night-

Backyard observers enjoy Comet Hale-Bopp during its visit to the Earth's vicinity in 1997. Thousands of new astronomy buffs were hatched as they watched the great comet—the brightest evening-sky comet visible from midnorthern latitudes since Halley's Comet in 1910.

sky observers who are unable to identify a single constellation. By using these two star groups, it is possible to identify every major star and constellation visible from Canada, the United States and Europe. Once Orion and the Big Dipper become familiar sights, the rest of the starry sky falls into place, no matter what the season or time of night.

Sky Motions

The stars move relative to one another, but the motion is so slight that the constellations remain intact for millennia. However, the sky as a whole has an apparent motion caused by the Earth's spinning on its axis (which produces day and night) and by the Earth's annual orbit around the Sun (our year). The daily axis rotation means that at different times of the night, any one point on Earth faces in different directions. You can observe this

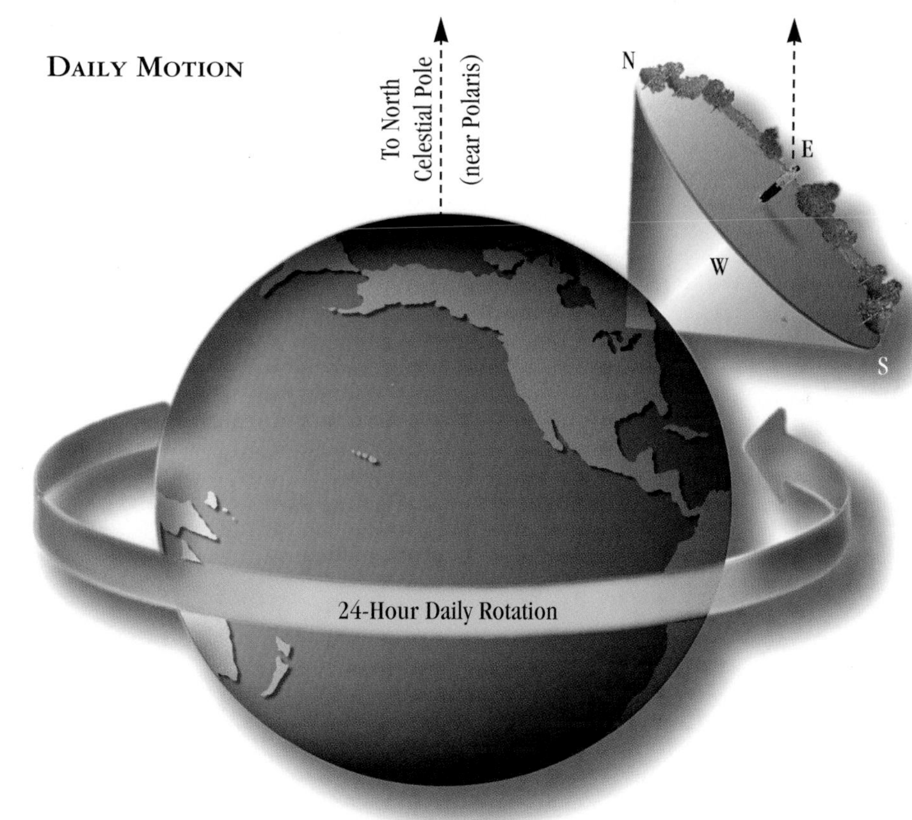

The daily rotation of Earth on its axis, right, carries us like passengers on a giant carousel. This motion causes the Sun to appear to rise and set each day and sweeps the stars across the night sky in great curving arcs, as seen in the time exposure above.

motion on a clear night by standing so that a bright star is just above a marker object—a pole or a house gable, for example. Note the position, and return to that spot 15 to 45 minutes later. The star will have moved a noticeable amount relative to that spot. In general, the stars swing from east to west, just as the Sun does each day. Stars in the north slowly twirl around the north celestial pole—the sky's pivot point—not far from Polaris, which hardly moves at all. The effect is vividly illustrated in the time-exposure photographs on these two pages.

The change of stellar scenery caused by the Earth's orbital trek around the Sun becomes apparent only over weeks or months, but the consequences are more profound. Whole sectors of the sky parade into view, changing just as the seasons do. Each season, therefore, has its prominent stars and constellations, which are illustrated in the following chapter.

Throughout this book and in general astronomy usage, the stars and constellations referred to in each season are those that the *evening* side of Earth is pointed toward. Since half of the entire sky is seen at any one time (Earth itself blocks the other half), some overlap occurs. In the spring sky, for example, a few winter and summer stars are also seen. By staying up all night, an observer gets a jump on the seasons, at least in a celestial context. As Earth rotates on its axis (counterclockwise, as viewed from the north, the same direction as its orbital revolution), the stars of the following season begin to appear in the east starting around midnight. By 4 a.m., the

ANNUAL MOTION

AUTUMN CONSTELLATIONS

SUMMER CONSTELLATIONS

Earth's Orbit

View is from above the Earth's North Pole

Sun

Rotation of Earth on its axis is counterclockwise

WINTER CONSTELLATIONS

SPRING CONSTELLATIONS

The nightly wheeling of the starry sky, top left, is a result of the Earth's daily axis rotation. A time-exposure star-trail photograph, above, demonstrates that motion in the southeastern sky. The camera was stationary for the 1-hour exposure, and the rotating Earth produced the star trails. Left: A second, much more ponderous sky motion is the annual revolution of Earth around the Sun, which causes the seasonal march of the constellations as we face different parts of our galaxy.

Sky Measures

Just as road maps have distance indicators between cities, our celestial guide maps denote distances between key stars and star groups—not the distance from Earth to the stars but, rather, the *apparent* distance from one star to another. This measure is calibrated in degrees (360 degrees in a circle). Using this calibration on the sky is beautifully simple: just hold up your hand. At arm's length, the width of the end of the little finger is almost exactly 1 degree—wide enough to cover the Sun or the Moon, both about half a degree across. The two pointer stars in the bowl of the Big Dipper used to find Polaris are 5 degrees apart, the width of three fingers held boy-scout fashion at arm's length.

For larger sky angles, one fist width is 10 degrees, while 15 degrees is the span between the first and little fingers spread out. An entire hand span, from thumb to little finger, is about 25 degrees, the length of the Big Dipper. Larger dimensions can be measured in multiples of these. For general reference, the distance from the horizon to overhead is 90 degrees. Remember, these hand-reference measurements work only at arm's length.

The system is reasonably accurate for men, women and children, since people with smaller hands tend to have shorter arms. Only the hand-span measure seems to vary from person to person, because some people can extend their thumb and little finger more widely than others can. A quick check against the Big Dipper will indicate whether you have a span closer to 20 degrees than 25. Anyone can become proficient at gauging the distances in degrees from one star or star group to another in just minutes.

It doesn't matter in which season you begin; the Big Dipper diagram on page 34 can be used to locate several prominent stars almost instantly once you have a sense of the dimensions involved.

This is the crucial first step toward becoming a backyard astronomer. Orion's seven brightest stars—three in the belt and four in a surrounding quadrilateral —are equally efficient as celestial guideposts. Orion's only drawback, compared with the Big Dipper, is that it is prominent in the evening sky only from late November to early April.

Backyard astronomy does not have to be a maze of formulas, calculators, grid lines, nomenclature, mythology and jargon. It can be easy and fun to find your way around the night sky. Most people want to be able to start finding celestial objects from their first night out. That's my goal here in laying out the most straightforward way to do it. In the next chapter, more detailed charts build on the same principle of using distinctive stellar guideposts to lead the observer around the sky. This is a gradual, painless way to come to know the starry night.

Polaris
28°
Big Dipper
25°
15°
10°
5°

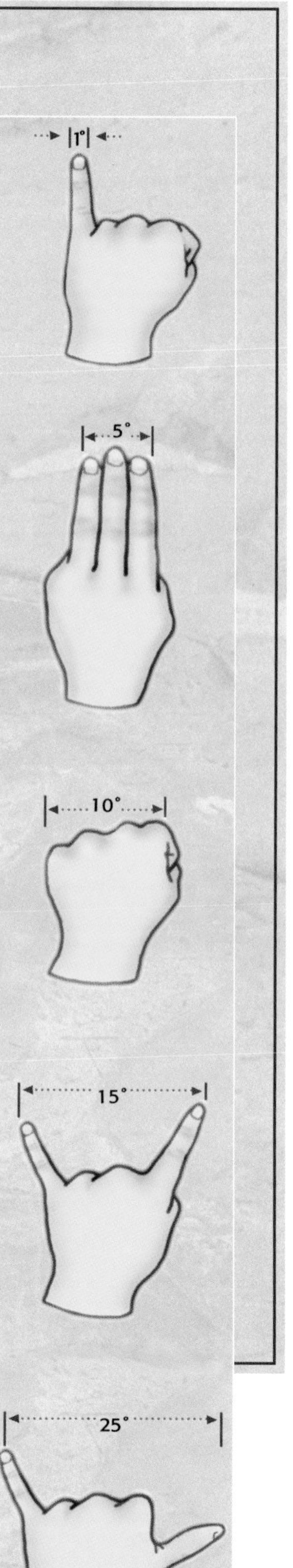

Finding your way around the night sky becomes a lot easier once you master this system of measuring off degrees. You can check your own hand measures against the Big Dipper stars. For instance, many people cannot spread their hand to achieve a full 25 degrees from little finger to thumb. You may reach only 20 or 22 degrees. But when you know your personal hand/sky measures, the system becomes amazingly accurate and endlessly useful.

observer has been transported one whole season ahead —that is, Earth has rotated to face the part of the sky seen in the evening hours of the following season.

Big Dipper Signpost

Although their positioning is random, the stars in the Big Dipper are, by some fortuitous cosmic coincidence, arranged in a pattern which provides at least seven sky pointers that guide the eye naturally to a dozen bright stars or stellar groups. The Big Dipper's prime importance as a key to the night sky is its constant availability for midnorthern-hemisphere observing. No matter what day of the year or what time of night, the Big Dipper is seen in the northern part of the sky every clear night from everywhere in Canada and from north of 40 degrees latitude in the United States. (Between 40 and 25 degrees north latitude, the Big Dipper is near or below the northern horizon for several weeks in autumn.) Orion in the southern sector of the sky augments the Big Dipper from November to April.

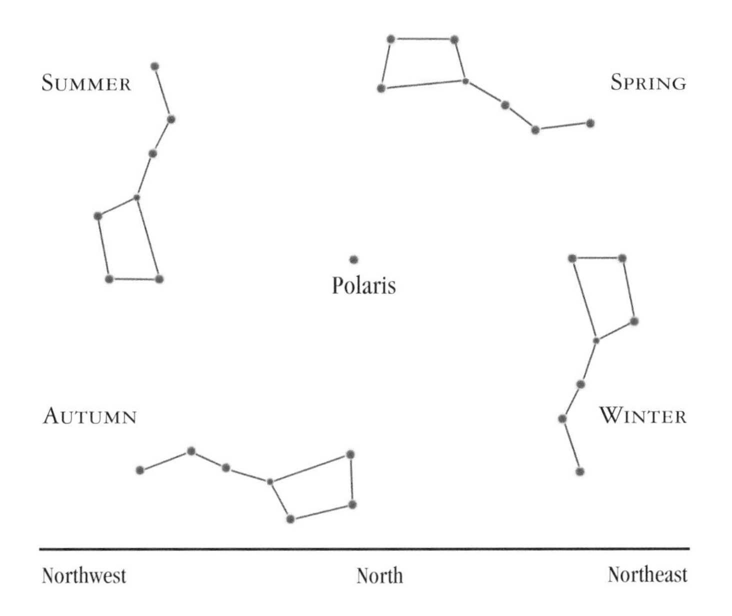

To locate these guiding constellations, consult the table on page 34. Although the Big Dipper and Orion move around the sky in a systematic way, so does everything else. The only additional information necessary before making the first series of star identifications is how to measure sky distances (facing page).

For stargazers living at midnorthern latitudes (Canada, the United States, Europe, Japan), the Big Dipper is the key to the night sky. Once you become familiar with its shape, you can track the Dipper's changing orientation during the year caused by the Earth's annual march in its orbit around the Sun. The photograph above shows the Big Dipper and Polaris in winter at the latitude of Boston or Chicago.

A Universe of Suns

Like our Sun, other stars produce their own light and energy through thermonuclear fusion reactions, the same process that generates the fury of a hydrogen bomb. Some stars are larger than the Sun; some are hotter. But all appear as points of light because of their enormous distances, the nearest being about 250,000 times farther away than our Sun.

Of the stars visible to the unaided eye, 99 percent are bigger and brighter than our Sun. This gives a distorted picture of the true variety of suns in space, because the majority of stars in the

Milky Way Galaxy—95 percent of them—are actually less luminous than the Sun, their comparatively feeble light output rendering them invisible to the backyard stargazer. Those seen with the unaided eye are the titans of the galaxy, the searchlights among throngs of 100-watt light bulbs.

The remoteness of these celestial beacons has another important consequence: stars do not appear to move relative to one another from week to week or even from one year to the next. Although they are traveling in space, that movement is insignificant compared with the distances between them. Our great-grandparents saw the stars of the Big Dipper exactly as we see them today. Star charts made by the Greek astronomer Hipparchus more than 2,000 years ago show the same stars in almost exactly the same positions as we now see them. (I say "almost" because one bright star and a handful of fainter ones have shifted about the width of the full Moon since then. But most have not altered their positions relative to each other by more than the width of a human hair held at arm's length.)

Star Brightness

A glance at the Big Dipper reveals that its stars are not the brightest in the sky. There are several stars substantially brighter and many significantly fainter. In the second century B.C., the Greek astronomer Hipparchus hit upon the idea of classifying the stars by their brightness. He decided to divide the stars into six categories. He designated the brightest as first magnitude and the faintest as sixth magnitude, with the others scattered in between.

This system is still in use today, although it has been refined and expanded to include telescopic stars fainter than sixth magnitude and objects brighter than first magnitude. A step of one magnitude is an increase or decrease by a factor of 2½ times in brightness. Therefore, an average first-magnitude star is 2½ times brighter than an average second-magnitude star, 6 times brighter than third, 16 times brighter than fourth, 40 times brighter than fifth and 100 times as brilliant as a sixth-magnitude star, the faintest that can be seen with the unaided eye on a very clear night.

No system is perfect. Some stars designated by Hipparchus as first magnitude are too bright. They are now rated zero; those brighter still are –1, and so on. (The scale is like an upside-down thermometer.) The magnitude classes extend in the faint direction all the way to 30th magnitude, the dimmest objects that can be detected with the Hubble Space Telescope. A sixth-magnitude star is four billion times brighter than one of 30th magnitude. At the opposite extreme, the brightness of the Sun is rated at magnitude –27, six trillion times brighter than a sixth-magnitude star.

Sirius, the brightest star in the night sky, is magnitude –1. Only Jupiter (–3), Venus (–4) and Mars (varies from +2 to –3) are brighter.

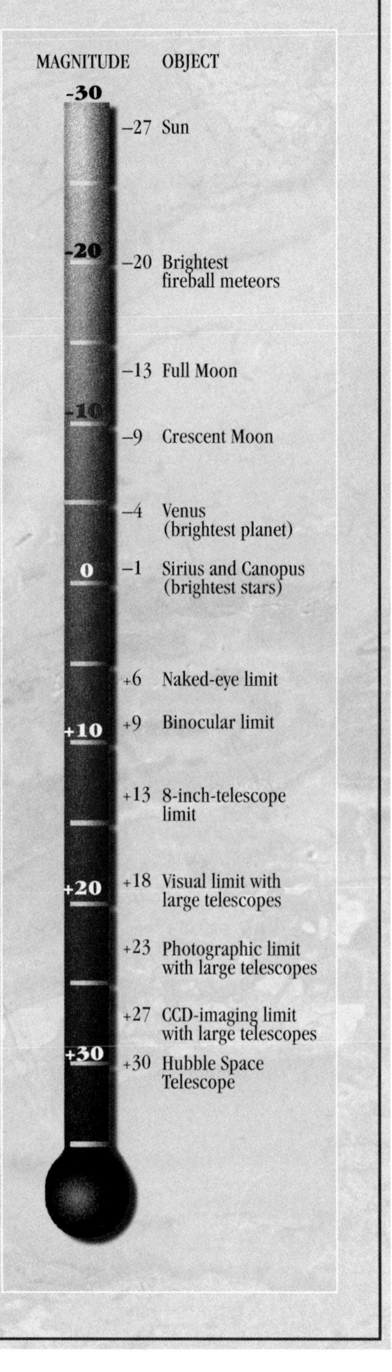

Lights that speckle the city at night demonstrate the reason that celestial objects appear bright or faint to the observer: some appear bright because they are nearby, while others are intrinsically brilliant but far more remote.

Constellations & Star Names

Long before Hipparchus, skywatchers of antiquity divided the sky into groups of stars called constellations. The stars that form a constellation are seldom related to one another. These celestial groupings are steeped in mythology and, in the case of the zodiac constellations, embroidered with the symbolism of astrology.

Today, astronomers still use the traditional Latin forms of the constellation names from classical Greek civilization. A few constellations (mostly dim ones) were invented in the 17th and 18th centuries to fill in sky regions not included in the ancient lore. In 1930, the constellation names and boundaries were officially set by the International Astronomical Union, and there have been no changes since. Although there are 88 constellations, one-quarter of them are in the southern sky, concealed from view in mid-northern latitudes; half of the remainder are rather faint. Being able to identify the dimmer constellations is not necessary in the beginning stages of amateur astronomy. Initially, you want to become familiar with the 15 or 20 brightest constellations.

The situation with individual star names is less formal than for constellations. Several hundred of the brightest stars have been named over the centuries, but only about 75 of these designations have survived the dustbin of disuse. As far as we know, the Babylonians were the first to name stars, although most of the names used today are Arabic, with a sprinkling of Greek, Latin and Persian. The preponderance of Arabic star names stems from the fact that

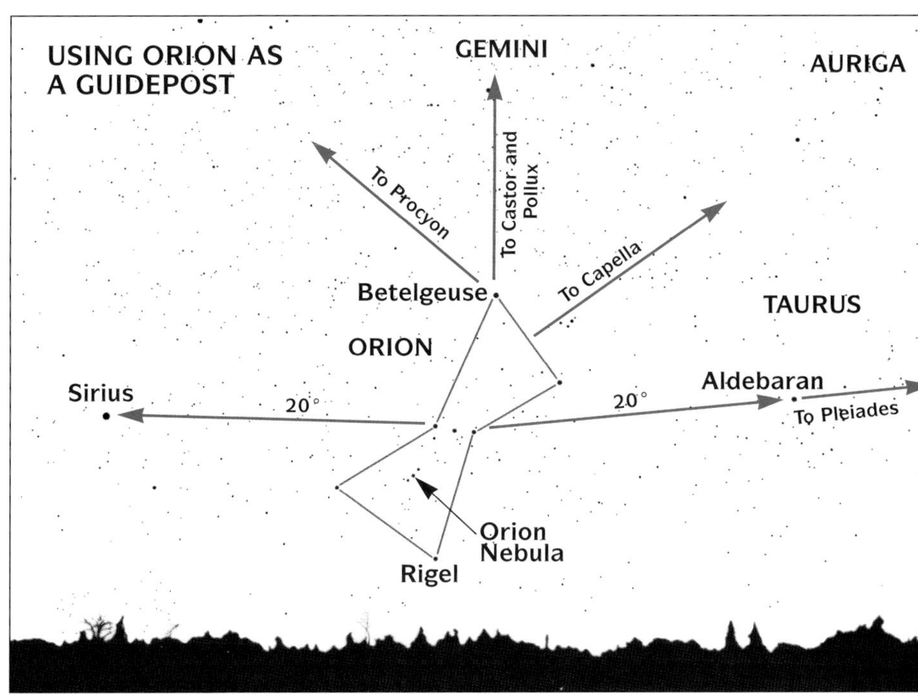

The starry night is steeped in myth and legend dating back thousands of years, left. The colorful tales of the constellations bring a rich history to astronomy but are of little aid to the beginning stargazer seeking to find his or her way around the night sky. Instead, the system used in this book focuses on two key starting points—the Big Dipper and Orion—which will guide the observer to all the major constellations. Above: Orion's seven bright stars, arrayed in a distinctive hourglass configuration, lead the eye to all the important stars and constellations of the winter sky. Orion stands above the western horizon, as seen here, in the evening sky in March and April.

during the Dark Ages, Arabic astronomy was the most advanced in the world. The Arab astronomers retained the tradition of Greek-Latin constellation names, but their star names superseded most earlier designations. Although many star names are meaningless in English, they usually translate into a logical word picture. Betelgeuse, for example, is believed to be ancient Arabic for "armpit of the mighty one." Spica is "ear of wheat" in Latin; Procyon means "before the dog" in Greek. A list of star and constellation names, with their meanings and a pronunciation guide, appears on pages 36 and 37.

Modern designations of stellar configurations, such as the Big Dipper and the Summer Triangle, hold no official status. They are convenient guides for casual stargazers, nothing more.

The Big Dipper: Key to the Night Sky

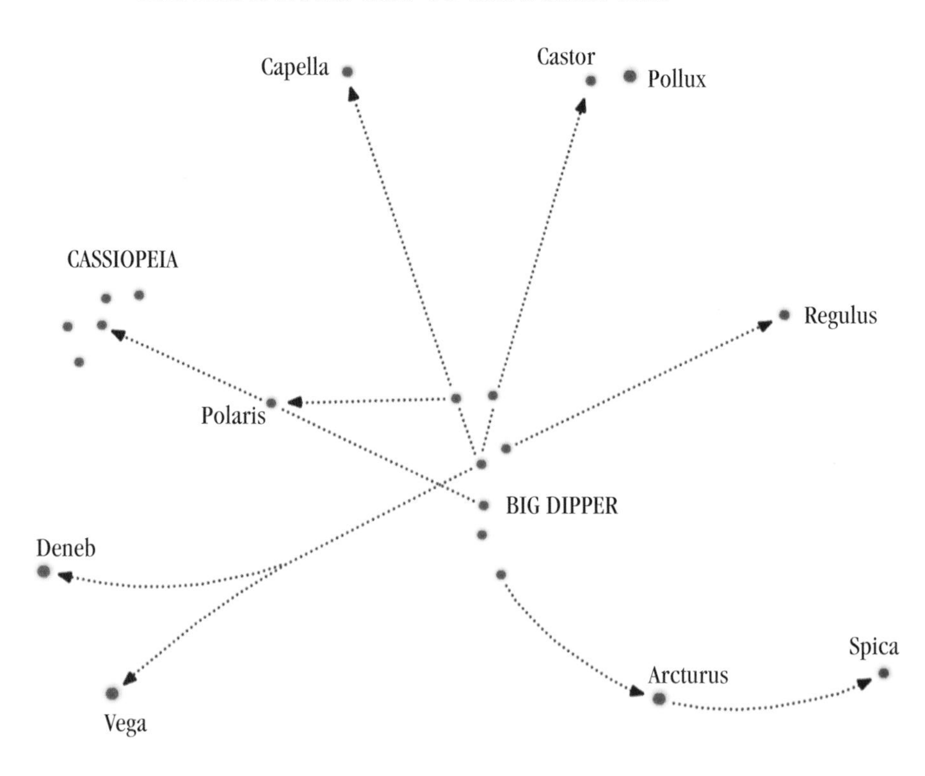

Where to Find the Big Dipper and Orion

Month	Big Dipper Direction	Big Dipper Altitude	Orion Direction	Orion Altitude
January	NE	25°	S	40°
February	NE	40°	S	45°
March	NE	55°	SW	35°
April	N	65°	SW	20°
May	N	70°	Not visible	
June	N	65°	Not visible	
July	NW	55°	Not visible	
August	NW	40°	Not visible	
September	NW	25°	Not visible	
October	N	15°	Not visible	
November	N	10°	E	15°
December	N	15°	SE	30°

By directly pointing the way to the brightest stars in nine different constellations, the Big Dipper is unrivaled as the premier signpost of the night sky. The diagram on this page should prove to be the most useful to anyone who has never before identified a star or a constellation. Keep in mind the differing orientation of the Big Dipper during the year (page 31). For practice, use the wide-angle photograph of the autumn sky above to identify Polaris, Cassiopeia, Vega, Deneb and Capella.

Information intended for midevening 8 to 10 p.m., local standard time, or 9 to 11 p.m., local daylight time; latitude 40 to 55 degrees N. In more southerly latitudes, the Big Dipper will appear proportionately lower and Orion higher in the sky.

Stuff That Moves

There is more to a starry night than just stars, planets and the Moon. Pulsating aircraft lights occasionally punctuate the scene. Darting flashes signal the incineration of meteors—bits of cosmic debris entering the Earth's atmosphere. And there are numerous moving lights that, apart from their motion, look like stars. These are Earth-orbiting satellites, shining with an unblinking white glow from sunlight reflecting off their metallic bodies and solar panels.

The prime time to look for satellites is during the first hour of darkness on spring and summer evenings. Try a reclining lawn chair, and watch the overhead region. Within a few minutes, you should see several starlike dots march through the constellations. One might be the space shuttle, another a secret military satellite, a third simply a spent rocket still in orbit. A satellite easily visible to the unaided eye is typically the size of a delivery van, travels at 28,000 kilometers per hour and crosses the sky in two to three minutes at an altitude of 300 to 500 kilometers.

Once you have a little experience, distinguishing the difference between a satellite and an airplane is easy. Most aircraft have either flashing lights or red or green wing lights, although a few have a steady white light like a satellite. Binoculars usually reveal engine exhaust or other lights on planes that appear to the naked eye as single white lights. Satellites always appear white, starlike and untwinkling. If the satellite disappears as it cruises across the sky, it has entered the Earth's shadow. The shadow climbs higher as the Sun sinks lower, which is why the best time to scan for satellites is the hour after darkness falls.

As a satellite passes over, it changes orientation with respect to the observer. Its brightness may surge for a few seconds when sunlight reflects off a solar panel or other flat surface. One group of satellites that takes this behavior to extremes is Iridium, a gang of 66 satellites used for global wireless-telephone communications. These space vehicles, each roughly the size of a small car, have flat, highly reflective telecommunications panels that are near-perfect reflectors. When one is lined up to reflect sunlight at Earth, a brilliant, slow-moving "star" appears that, within seconds, reaches a peak brightness as great as magnitude –7, then just as rapidly fades to obscurity. As of mid-1998, Iridium flash predictions for your location, along with other satellite viewing times, can be obtained at the following Website: http://www.gsoc.dlr.de/satvis/

Well away from city lights, a careful observer should see at least 10 satellites in that crucial first hour after nightfall. The number declines after that and is quite low around midnight. Sightings also drop off during the fall and winter months, when the Earth's shadow is higher in the sky. Sometimes, a satellite appears to pulsate in a regular rhythm, which means the entire device is tumbling. Active satellites are always stabilized, so if the pulsing is obvious, you can be sure that you're seeing something inactive or an old rocket body.

How much stuff is up there? The U.S. Air Force Space Command keeps tabs on about 8,000 objects, ranging from the bus-sized Russian Mir space station (below) to a hinge the size of a paperback book that broke away from a European satellite's covering in 1988. Fewer than 1,000 orbiting objects are doing something useful, the rest are rubbish that has accumulated since the dawn of the space age. Occasionally, a space shuttle's flight is altered to avoid a close brush with one of these potentially lethal chunks.

Whether satellites have a steady or a fluctuating brightness, most novice observers agree that they do not appear to move across the sky in a perfectly straight line. There seems to be a perceptible waviness to their paths or a jerkiness in speed as they glide through the starry background. In fact, these oscillations are in the mind, not the sky. The satellites actually move at a smooth pace in precise linear paths.

The human brain likes to link patterns into a recognizable image. This is done instantaneously in daily life. However, when we look at one moving light in a randomly dotted black sky, the brain constantly tries to produce these patterns but fails. What are thought to be oscillations in the satellite's path are really the unconscious workings of the mind trying to make sense out of an unfamiliar visual environment. The result is, in effect, an optical illusion.

Star & Constellation Pronunciation Guide

Name	Object	Pronunciation	Meaning*
Acamar	star in Eridanus	ACHE-uh-mar	end of river (A)
Achernar	brightest star in Eridanus	ACHE-er-nar	end of river (A)
Adhara	star in Canis Major	ah-DARE-rah	the maiden (A)
Albireo	star in Cygnus	al-BEER-ee-oh	[meaning unknown]
Alcor	star in Ursa Major	AL-core	the abject one (P)
Alcyone	brightest star in Pleiades	al-SIGH-oh-nee	one of the Pleiades sisters from mythology (G)
Aldebaran	brightest star in Taurus	al-DEB-uh-ran	the follower [of the Pleiades] (A)
Alderamin	star in Cepheus	al-DARE-uh-min	the right forearm (A)
Algieba	star in Leo	al-JEE-buh	the forehead (A)
Algenib	star in Pegasus	al-JEE-nib	the flank (A)
Algol	variable star in Perseus	AL-gall	the ghoul (A)
Alhena	star in Gemini	al-HE-na	the brand mark (A)
Alioth	star in Big Dipper	ALLEY-oth	the goat (A)
Alkaid	star in Big Dipper	al-KADE	daughter of the bear (A)
Almach	star in Andromeda	AL-mac	the weasel (A)
Alnair	brightest star in Grus	al-NAIR	the bright (A)
Alnilam	star in Orion's belt	al-NIGH-lam	the arrangement [of pearls] (A)
Alnitak	star in Orion's belt	al-NIGH-tak	the belt (A)
Alpha Centauri	brightest star in Centaurus	AL-fah sen-TORE-eye	[modern designation]
Alphard	brightest star in Hydra	AL-fard	the solitary (A)
Alphecca	brightest star in Corona Borealis	al-FECK-ah	the broken ring [of stars] (A)
Alpheratz	star in Andromeda	al-FEE-rats	navel of the steed (A)
Altair	brightest star in Aquila	al-TAIR	the flying one (A)
Andromeda	prominent constellation	an-DROM-eh-duh	daughter of Cassiopeia in mythology (G)
Antares	brightest star in Scorpius	an-TAIR-eez	rival of Mars (G)
Aquarius	zodiac constellation	a-QUAIR-ee-us	the water carrier (L)
Aquila	prominent constellation	A-quill-ah	the eagle (L)
Arcturus	brightest star in Bootes	ark-TOUR-us	bear guard (G)
Aries	zodiac constellation	AIR-eez	the ram (L)
Arneb	brightest star in Lepus	AR-neb	the hare (A)
Auriga	prominent constellation	oh-RYE-gah	the charioteer (L)
Bellatrix	star in Orion	bell-LAY-trix	the warrioress (L)
Betelgeuse	star in Orion	BET-el-jews	armpit of the mighty one (A)
Bootes	prominent constellation	bo-OH-teez	the herdsman (G)
Canes Venatici	small constellation	KAY-neez ve-NAT-ih-sigh	the hunting dogs (L)
Canis Major	prominent constellation	KAY-niss MAY-jer	the great dog (L)
Canis Minor	small constellation	KAY-niss MY-ner	the lesser dog (L)
Canopus	brightest star in Carina	can-OH-pus	the helmsman (G)
Capella	brightest star in Auriga	kah-PELL-ah	the she-goat (L)
Caph	star in Cassiopeia	kaf	the hand (A)
Carina	prominent southern constellation	ka-RYE-nah (or ka-REE-nah)	the keel [of the ship *Argo*] (L)
Cassiopeia	prominent constellation	KAS-ee-oh-PEE-ah	wife of Cepheus in mythology (G)
Castor	star in Gemini	KAS-ter	the beaver (G)
Centaurus	prominent southern constellation	sen-TOR-us	the centaur (G)
Cepheus	constellation	SEE-fee-us	king of Ethiopia in mythology (G)
Cetus	large, dim constellation	SEE-tus	the whale menacing Andromeda (G)
Coma Berenices	small constellation	KOH-mah bera-NICE-eez	Berenice's hair (G)
Cor Caroli	brightest star in Canes Venatici	kor CARE-oh-lie	heart of Charles [Charles II of England] (L)
Corona Borealis	small constellation	kor-OH-nah bo-ree-ALICE	the northern crown (L)
Corvus	small constellation	CORE-vus	the crow (L)
Cygnus	prominent constellation	SIG-nus	the swan (G & L)
Delphinus	small constellation	del-FINE-us	the dolphin (G & L)
Delta Cephei	variable star in Cepheus	DEL-ta SEE-fee-eye	[an important variable star]
Deneb	brightest star in Cygnus	DEN-eb	tail of the hen (A)
Denebola	star in Leo	duh-NEB-oh-lah	tail of the lion (A)
Diphda	brightest star in Cetus	DIFF-dah	the frog (A)
Draco	constellation	DRAY-ko	the dragon (G)
Dschubba	star in Scorpius	JEW-bah	the forehead (A)
Dubhe	star in Big Dipper	DUE-bee	the bear (A)
Elnath	star in Taurus	el-NATH	the butting [horn] (A)
Eltanin	star in Draco	el-TAY-nin	the sea monster (A)
Enif	star in Pegasus	EN-if	the nose [of the horse] (A)

Equuleus	small constellation	ee-KWOO-lee-us	the little horse (L)
Eridanus	constellation	eh-RID-an-us	a river (G)
Fomalhaut	brightest star in Piscis Austrinus	FOAM-a-lot (or FOAM-ah-low)	mouth of the fish (A)
Gemini	zodiac constellation	GEM-in-eye	the twins (G)
Hadar	star in Centaurus	HAD-ar	the settled land (A)
Hamal	brightest star in Aries	HAM-el	the ram (A)
Hyades	star cluster in Taurus	HI-a-deez	half-sisters to the Pleiades (G)
Izar	star in Bootes	EYES-ar	the loincloth (A)
Kochab	star in Ursa Minor	KOE-kab	the star (A)
Lacerta	small constellation	la-SIR-tah	the lizard (L)
Lepus	constellation	LEE-pus	the hare (L)
Libra	zodiac constellation	LYE-bra (or LEE-bra)	the balance (L)
Lupus	constellation	LEW-pus	the wolf (L)
Lyra	prominent constellation	LYE-rah	the lyre (G)
Markab	star in Pegasus	MAR-keb	the part for riding on (A)
Megrez	star in Big Dipper	ME-grez	the insertion point [of the bear's tail] (A)
Menkalinan	star in Auriga	men-KAL-in-an	shoulder of the charioteer (A)
Menkar	star in Cetus	MEN-kar	the nostril [of the whale] (A)
Menkent	star in Centaurus	MEN-kent	[modern corruption for Centaurus's shoulder]
Merak	star in Big Dipper`	ME-rac	the loins [of the bear] (A)
Mintaka	star in Orion's belt	min-TAK-uh	the belt (A)
Mira	variable star in Cetus	MY-rah	the wonderful (A)
Mirfak	brightest star in Perseus	MUR-fak	the elbow (A)
Mirzam	star in Canis Major	MUR-zam	the roarer [announcing Sirius] (A)
Mizar	star in Ursa Major	MY-zar	the wrapping (A)
Monoceros	constellation	mon-OSS-err-us	the unicorn (G)
Nunki	star in Sagittarius	NUN-key	Sumerian for god of the waters
Ophiuchus	constellation	oh-fee-YOU-kus	the serpent bearer (G)
Orion	prominent constellation	oh-RYE-un	the hunter (G)
Pegasus	prominent constellation	PEG-uh-sus	the winged horse (G)
Perseus	prominent constellation	PURR-see-us	hero; rescuer of Andromeda (G)
Phact	star in constellation Columba	fact	the dove (A)
Phecda	star in Big Dipper	FECK-duh	the thigh [of the big bear] (A)
Pisces	zodiac constellation	PIE-sees	the [two] fishes (L)
Piscis Austrinus	constellation	PIE-sis OSS-TRY-nus	the southern fish (L)
Pleiades	star cluster in Taurus	PLEE-ah-deez	the seven sisters (G)
Polaris	the North Star	poh-LAIR-iss	[north] pole star (L)
Pollux	brightest star in Gemini	PAW-lux	much wine (L)
Porrima	star in Virgo	poh-RIM-ah	goddess of childbirth (L)
Praesepe	star cluster in Cancer	pray-SEEP-ee	the manger (L)
Procyon	brightest star in Canis Minor	PRO-see-on	before the dog (G)
Rasalgethi	star in Hercules	RAS-el-GEE-thee	head of kneeling one (A)
Rasalhague	star in Ophiuchus	RAS-el-HAY-gwee	head of the snake man (A)
Regulus	brightest star in Leo	REGG-u-lus	prince (L)
Rigel	brightest star in Orion	RYE-jel	the foot (A)
Sabik	star in Ophiuchus	SAY-bik	the preceding (A)
Sadr	star in Cygnus	SAD-er	the breast [of the swan] (A)
Sagitta	small constellation	sah-JIT-ah	the arrow (L)
Sagittarius	prominent zodiac constellation	saj-ih-TAIR-ee-us	the archer (L)
Saiph	star in Orion	saw-EEF (or safe)	the sword (A)
Scheat	star in Pegasus	SHEE-at	the leg (A)
Schedar	star in Cassiopeia	SHED-ar	the breast (A)
Scorpius	prominent zodiac constellation	SKOR-pee-us	the scorpion (G)
Scutum	small constellation	SKEW-tum	the shield (L)
Shaula	star in Scorpius	SHOAL-ah	the raised [tail] (A)
Sirius	brightest star in Canis Major	SEAR-ee-us	scorching (G)
Spica	brightest star in Virgo	SPIKE-ah	ear of wheat [held by Virgo] (L)
Tarazed	star in Aquila	TAR-uh-zed	plundering falcon (P)
Taurus	prominent zodiac constellation	TOR-us	the bull (G)
Thuban	star in Draco	THEW-ban	the snake (A)
Vega	brightest star in Lyra	VAY-gah (or VEE-gah)	the stooping [eagle] (A)
Virgo	prominent zodiac constellation	VURR-go	the maiden (L)
Vulpecula	small constellation	vul-PECK-you-lah	the fox (L)
Zubenelgenubi	star in Libra	zoo-ben-ell-jen-NEW-bee	the southern claw (A)
Zubeneschamali	star in Libra	zoo-ben-ess-sha-MAY-lee	the northern claw (A)

*A = Arabic, G = Greek, L = Latin, P = Persian; derivation of meaning is sometimes a very rough translation

CHAPTER FOUR

STARS FOR ALL SEASONS

Who were they, what lonely men, Imposed on the fact of night,
The fiction of the constellations?

Patric Dickinson

Besides being a pleasant way to pass a mild evening under the night sky, learning to identify the stars and constellations is the foundation for all other elements of exploring the universe from your backyard. The nightly canopy of stars is a celestial map with which the observer must be familiar before seeking out specific targets for binocular and telescope viewing. The essentials of star identification were outlined in the previous chapter. Now comes the integration to the complete night sky.

All too often, sky charts intended for beginners sacrifice realism and clarity by adding celestial grids, telescopic objects and constellation and star names. This book utilizes a unique all-sky dual-chart system. Full-color charts for each season, showing the stars as realistically as possible, are paired with identical charts that include names and the complete locater-arrow system introduced in Chapter 3. When used together, these sky charts surmount many of the problems inherent in sky charts of the past. (More detailed charts are introduced in Chapter 6.)

The all-sky color charts are a miniature planetarium in a book. They duplicate the appearance of the night sky seen from a typical dark, but not necessarily pitch-black, location in southern Canada, the United States (except extreme southern Florida and Texas) and most of Europe. The locater-arrow system itself will work anywhere in the northern hemisphere.

Each chart is keyed to the evening hours of a particular season, but on almost any night of the year, two of the charts will be usable—one for the evening hours and one for the early morning. Presenting the entire visible sky in one illustration allows rapid linking of one star group to another, and gradually, the night sky will become woven into a single mental picture.

In essence, the all-sky seasonal charts reproduce the dome of the night sky on a flat surface. Thus the horizon becomes

Orion the mighty hunter strides above the Rocky Mountains in this winter scene from Lake Louise, Alberta.

Facing page: Cassiopeia's distinctive W-shape is a guide to the autumn night sky.

the edge of the chart, and the overhead point is at the center.

The charts are most practical if used a section at a time. Human eyes are not capable of taking in the entire sky at once anyway. Normally, only about one-quarter of the whole celestial dome can be comfortably viewed without substantial head movement. In order to make the chart conform to any quadrant of the sky, the book must be turned so that the compass point the observer is facing is down. When using the spring chart and facing east, for example, the book should be rotated 90 degrees counterclockwise so that the east compass point is at the bottom. The curving horizon line then corresponds to the actual horizon and looks like the illustration on facing page. The identities of the stars and constellations in this same region are supplied on the black-and-white chart.

If you move around the horizon a wedge at a time, the big picture should soon come into focus. For each season, there is a preferred starting point involving the Big Dipper or Orion. A step-by-step approach is described later in this chapter, but first, let's consider the charts.

The All-Sky Charts

All stars down to third magnitude and many to fourth magnitude are shown on each seasonal all-sky chart. The charts would become a bewildering maze of dots if fifth- and sixth-magnitude stars were included. (However, the full sky to fifth magnitude is detailed in the series of 20 charts in Chapter 6.) Here are some suggestions for getting the most use out of the seasonal all-sky charts:

1. Although the charts are designed for specific timespans, they are still useful for up to one hour on either side of the intervals indicated, except for locating objects near the horizon.

2. The book's spiral binding allows the facing all-sky charts to be folded back-to-back for outdoor use. One side is the real-sky simulation; the other identifies the stars and constellations.

3. When starting out, avoid hazy skies or nights washed by the full Moon's light. Too few stars will be visible on these nights for proper identification. Conversely, pitch-black skies, although inspiring, sometimes reveal such a profusion of stars that initial identification is difficult.

4. If at all possible, select an observing site that is protected from a direct view of yard lights or streetlights. You may have to sacrifice sections of the sky to do this by positioning yourself to get a house, a hedge or some other obstruction to block the offending glare, but the stars you do see will be more obvious, because your eyes will have an opportunity to become adapted to the darkness.

Essential Information for Using the All-Sky Charts Outdoors

• The edge of the chart represents the horizon; the overhead point is at center.

• The chart is most effective when you use about one-quarter of it at a time, which roughly equals a comfortable field of view in a given direction.

• To use the chart, hold it in front of you, with the chart rotated so that the direction you are facing is at the bottom. Don't be confused by the east and west points on the chart lying opposite their location on a map of the Earth. When held as shown here, the chart directions will match the compass points.

• On a moonless night in the country, you will see more stars than are shown on the chart; deep in the city or around full Moon, you will see fewer.

• The ecliptic is the celestial pathway of the Moon and planets. The star groups straddling this line are known as the zodiac constellations.

• For the best results when reading the chart outdoors, use a flashlight heavily dimmed with red plastic. Unfiltered lights greatly reduce night-vision sensitivity.

The key to effective use of the all-sky charts is to rotate the chart so that the direction you are facing is at the bottom. Always illuminate the chart with a red-filtered flashlight to preserve your eyes' dark adaptation.

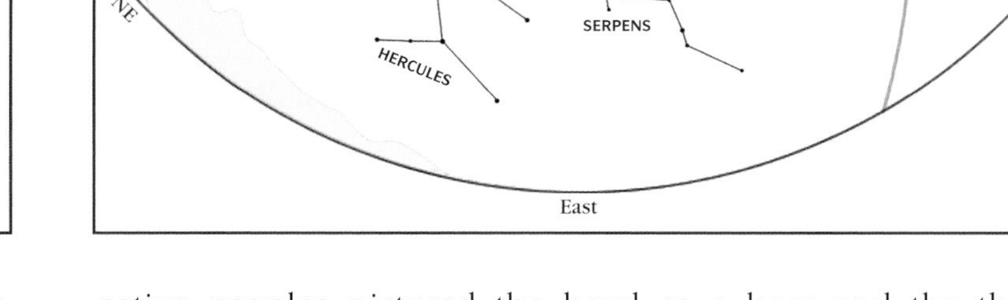

5. Work from the known to the unknown. Start with the Big Dipper, and utilize the locater arrows. Be patient—it usually requires at least a year to become completely comfortable with star identification.

6. If a bright star is seen in the vicinity of the ecliptic line, it is almost certainly a planet. (The ecliptic is the celestial pathway of the Sun, Moon and planets.) Sorting out the identities of the five naked-eye planets is described in Chapter 7.

7. When using this book outdoors, illuminate the pages with a flashlight heavily filtered with red plastic or cellophane. If nothing else is available, use several layers of brown paper to dim the flashlight. Turning on an unfiltered flashlight to see the chart will destroy your eyes' sensitivity to the darkness, and it will take several minutes for them to recover.

8. Dim indoor lights before going outside. Your eyes will become sensitive to low light levels much more quickly, allowing fainter stars to be seen sooner. While outside, avoid looking at streetlights and house lights as much as possible. Not only does direct artificial illumination spoil the aesthetic appeal of the sky, but such lighting also affects the eyes' sensitivity to the dark. (For more on this, see page 84.)

The Spring Sky

My normal enthusiasm for astronomy always gets an extra boost on that first mild spring evening when the stars shine and conditions bode well for a long season of skygazing. The Big Dipper is nearly overhead throughout spring evenings, its elaborate system of pointer stars providing the best possible opportunity to link up all the major stars and star groups above the horizon.

The Big Dipper is not a true constellation. Rather, it is the brightest part of the sprawling constellation Ursa Major, the large bear in mythology that guards the polar regions. The Dipper is the invention of 19th-century American stargazers. In Britain, the seven Dipper stars are known as the Plough. North American native peoples pictured the bowl as a bear and the three handle stars as a trio of braves stalking the beast.

A mental extension of the curve of the Big Dipper's handle one full Dipper length reaches zero-magnitude Arcturus, the brightest star in the spring skies. Arcturus is the most prominent star in the constellation Bootes the herdsman. Its name and location can be memorized with the phrase "follow the arc to Arcturus," which refers to the arcing curve made by extending the Big Dipper's handle. Sometimes added to this is "and speed on to Spica," which can be done easily by extending the curve another Dipper length to the first-magnitude star Spica, in the large zodiac constellation Virgo. And the curve does not end there. Another 15-degree extension leads to a small but conspicuous quadrilateral of third-magnitude stars known as Corvus. Corvus's identity can be confirmed by using its top two stars as pointers back to Spica.

The two stars in the Big Dipper's bowl nearest the handle can be used to form a locater arrow running 45 degrees south to Regulus, the first-magnitude star in Leo the lion. A backward question mark signifies the beast's head and mane, while Regulus is Leo's heart. His hindquarters are designated by a triangle of stars to the east. A faint chain of stars meandering below the triangle is the lion's tail. Overall, the stars of Leo cover an area of sky slightly larger than the Big Dipper. Leo is the most prominent of the spring constellations and the only one in the spring sky that resembles the object for which it was named.

The all-sky charts in this chapter are most effective when used sectionally—that is, just use the part above the horizon you are facing. With paired charts back to back, you can flip the book over to alternate between guide chart and full-color chart.

The locater arrow that traces diagonally across the Dipper's bowl to Castor and Pollux in Gemini is an important link between spring and winter constellations. Grouping constellations by seasons is a convenience based on when they are most prominently visible in the evening sky. The annual march of the seasons is reflected in the sky by a procession of constellations. In spring, the night side of Earth faces the region in space decorated by the stars of Leo, Bootes and Virgo. The winter groups are seen low in the west, dipping into the twilight glow. By late spring, they are invisible, lost in the Sun's glare as Earth continues around its orbit.

Of the three brightest stars in spring skies, Regulus and Spica are first magnitude and Arcturus is zero magnitude. Regulus is a bluish star (although it appears white to the eye) about 78 light-years distant, with a luminosity roughly 150 times that of the Sun. Spica is actually 10 times as bright as Regulus and 4 times as remote.

Arcturus, only 37 light-years away, is one of the nearest bright stars. It's a giant, about 23 times the diameter of our Sun and radiating 130 times as much energy. Its pale orange color is evident even to the unaided eye.

Choose a dark night in spring to look for a beautiful sprinkling of stars about midway between Regulus and Bootes. Called Coma Berenices, this little constellation (not shown on these charts) is a star cluster—a group of stars born at approximately the same time and still grouped together in space. Coma Berenices is some 250 light-years distant, which makes it the nearest star cluster after the Hyades, in the constellation Taurus. Fewer than a dozen of the cluster stars can be seen with the unaided eye. But if binoculars are turned on Coma Berenices, a dozen or two more are splashed against a memorable array of background stars.

The Beehive star cluster (M44), located midway between Regulus and Pollux, is twice the distance of Coma Berenices. Richer and more compact, the Beehive—a pale fuzz to the naked eye—is a prize binocular target. Look for it on the color chart as a light smudge, just as it appears to the unaided eye.

Binoculars will also reveal the star Alcor, a companion to Mizar, the star at the bend of the Big Dipper's handle. The two travel together through space, about three light-years apart. Alcor can be spotted without optical aid, but binoculars make it easy. (Many more binocular objects are plotted and described on the charts in Chapter 6.)

A lone binocular observer enjoys the celestial sights on a summer evening in the country. The brightest object is Jupiter, which was in the constellation Sagittarius at the time.

The Summer Sky

If astronomy is good for the soul, then summer is the time for meditation. Summer nights away from the city offer the prime opportunity for discovering the stars. Summer stars in rural skies are so bright that they somehow seem closer, more accessible. With the tranquillity of the night and the majesty of the stellar array, stargazing becomes an almost hypnotic experience, like watching dancing flames in a fireplace.

Exploring summer skies is intimately linked to the Summer Triangle, a large, distinctive figure that is strictly a modern invention. The triangle is Vega, Deneb and Altair, the brightest stars of three separate constellations. They are so much brighter than their neighbors that the triangle is the dominant summer and early-autumn configuration.

To find it, return to the great stellar signpost, the Big Dipper, now in the northwest. Extend a line out the open end of the bowl from the two bowl stars nearest the handle to a point midway between Vega and Deneb, about 60 degrees away. The two are quite easy to distinguish from one another: Vega, at zero magnitude, is noticeably brighter than first-magnitude Deneb. The triangle is completed by Altair, which is also first magnitude but somewhat brighter than Deneb. The Summer Triangle covers a huge patch of sky, larger than the area masked by an open hand held at arm's length.

The most prominent of the constellations associated with the Summer Triangle is Cygnus, its main stars forming a cross, with Deneb at its top. Popularly known as the Northern Cross, Cygnus is a swan in mythology, its tail at Deneb and its wings stretching beyond the cross's arms. The swan's neck extends to the foot of the cross, the third-magnitude star Albireo. Vega's constellation is the small but distinctive Lyra the harp. Altair is the brightest star in Aquila the eagle, a collection of

The Galaxy in a Sandbox

A sandbox is a miniature universe for a child, a world of valleys, castles, hideaways and great escapes. On a different level, a sandbox serves as a model of the universe. There are about the same number of sand grains in a typical sandbox as there are stars in the Milky Way Galaxy.

When examined in a microscope, all the grains are seen to be fundamentally the same silicate material but are different in detail from each other. So it is with the stars. They are all cosmic thermonuclear furnaces, but they differ in size, temperature and brightness. Our Sun is one of them.

Just as the sand grains merge into a smooth texture to the eye, the stars blend into a radiant beach head, seen as the Milky Way on summer nights. Only the nearest stars, those within a few thousand light-years of the Sun, are visible as individuals. A thimbleful of sand from the box would represent all the stars visible to the unaided eye on a dark night.

But the sandbox is just our galaxy, and ours is merely one of billions. Every human on Earth would need a sandbox to begin to approach a representation of the real universe, and even that would fall short by several billion sandboxes. Just counting the sandboxes—the galaxies in the known universe—at the rate of one per second, 24 hours a day, would take several human lifetimes.

I am often asked whether the immensity of the universe makes me feel totally insignificant, even depressed, when I am out stargazing. On the contrary, I feel a deep sense of tranquillity under the starry night sky. It's not an unfathomable mystery but a wonderland to be explored. Humans may not understand all the intricate workings of the universe, but we do know enough to recognize our place in the cosmic scheme—at least in a physical sense. For that reason alone, we are not so insignificant.

Stargazing for me is a cerebral voyage among the stars and galaxies, a communion with the beauty and immensity of the universe. It isn't overwhelming—it's exhilarating.

Those feelings are reinforced every time I stand under a rich canopy of stars and see the Milky Way's glowing spine of starlight arcing across the sky. I sink back in a lawn chair and turn my binoculars to the throngs of stellar points in Cygnus and Sagittarius. The ocean of stars offered by humble binoculars never loses its impact.

The late Canadian astronomer Helen Hogg summed it up in an essay she wrote in the 1970s: "Many people tend to postpone their enjoyment of the stars because they are constantly with us, but...once you come to know [the stars], they never lose their appeal."

The spiral galaxy M101, although somewhat larger than the Milky Way, is in many ways a near twin of our home galaxy. If the Sun were located in one of the spiral arms of M101, our night sky would probably look much the same.

Spring

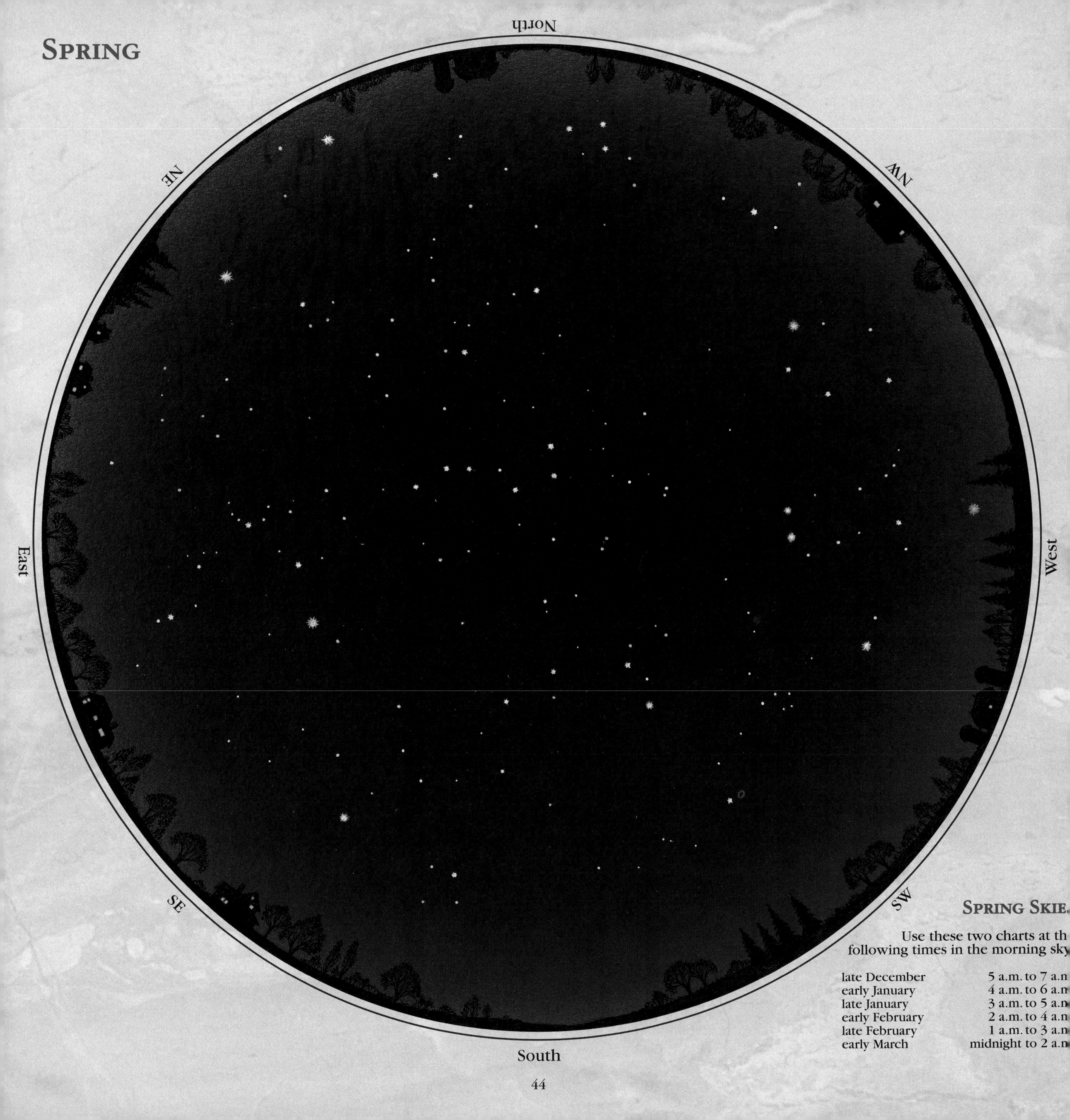

Spring Skie

Use these two charts at th
following times in the morning sky

late December	5 a.m. to 7 a.n
early January	4 a.m. to 6 a.n
late January	3 a.m. to 5 a.n
early February	2 a.m. to 4 a.n
late February	1 a.m. to 3 a.n
early March	midnight to 2 a.n

Spring

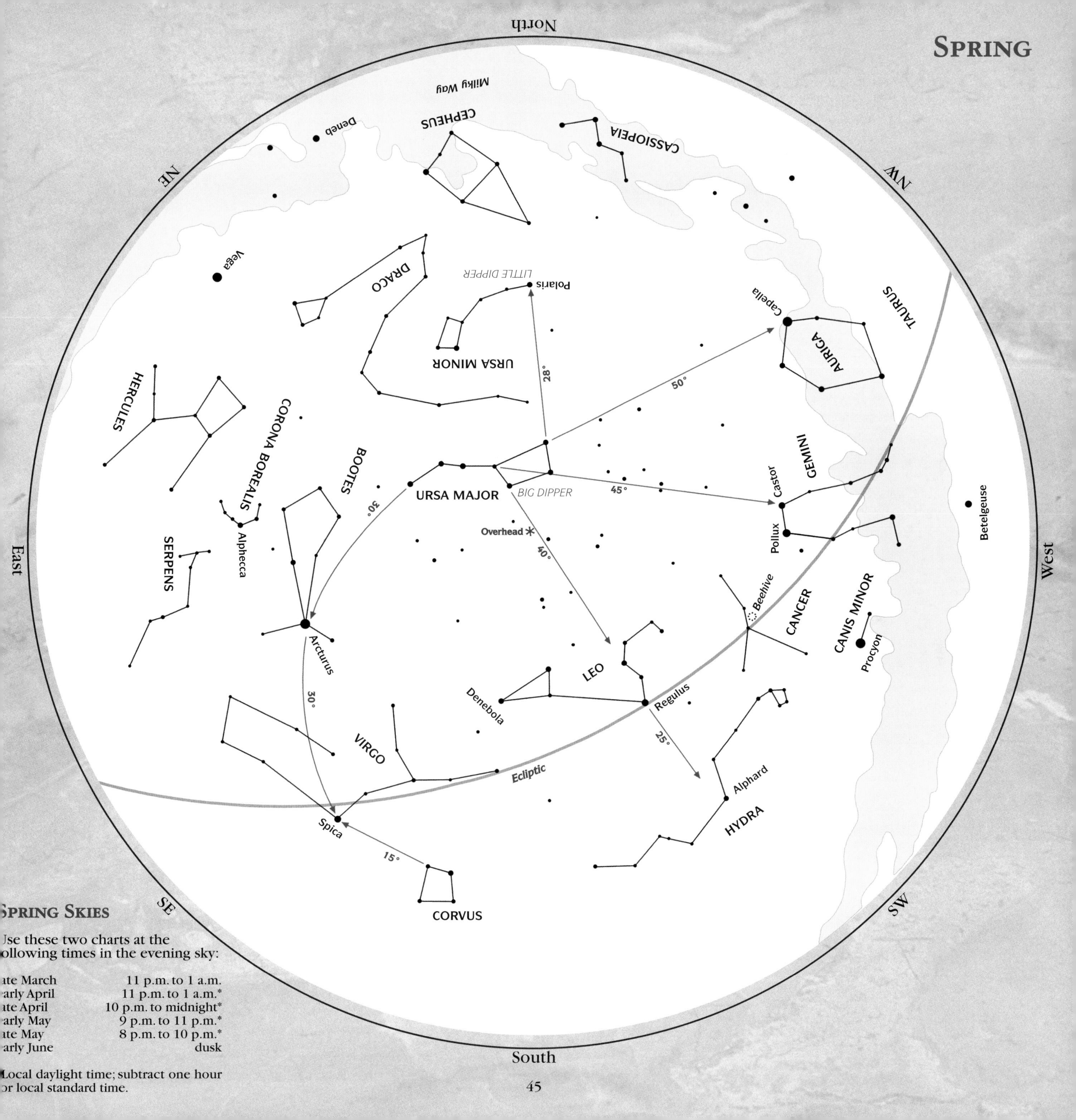

Spring Skies

Jse these two charts at the ollowing times in the evening sky:

ate March	11 p.m. to 1 a.m.
arly April	11 p.m. to 1 a.m.*
ate April	10 p.m. to midnight*
arly May	9 p.m. to 11 p.m.*
ate May	8 p.m. to 10 p.m.*
arly June	dusk

Local daylight time; subtract one hour or local standard time.

third- and fourth-magnitude stars with a vaguely birdlike outline.

With the Summer Triangle's three stars comfortably verified, extend a sight line from Vega to equally bright Arcturus in the west, the major bright star of spring skies. (Arcturus can be confirmed by using the locater arrow that curves outward from the Big Dipper's handle.) This Vega-Arcturus line passes directly through the constellations Hercules and Corona Borealis, the northern crown. Corona Borealis is a small but conspicuous arc of third- and fourth-magnitude stars set off by second-magnitude Alphecca.

The stars of Hercules are more dispersed and less easily distinguished. The Vega-Arcturus line passes through a quadrilateral composed of third- and fourth-magnitude stars, probably the constellation's most distinctive feature. Hercules is the sky's fifth largest constellation in terms of its officially allotted area in square degrees—only Hydra, Virgo, Ursa Major and Cetus are larger—yet its sprawling territory does not include a single star brighter than third magnitude.

Hercules is not alone in being both large and faint. Sometimes, a constellation has so few bright stars that the region in which it is located seems barren. Such is the case with Ophiuchus, a group whose brightest star, Rasalhague, is about the same magnitude as the central star in the Northern Cross. A direction line running 50 degrees from Deneb through the Summer Triangle, just skimming the southern end of Lyra, leads to Rasalhague. Misidentification is unlikely, since Rasalhague is the brightest star between the Summer Triangle and Scorpius, low in the south. Ophiuchus has all its reasonably bright stars at its periphery, leaving a vast blank zone.

The brightest star in the southern summer sky is Antares, located in the fishhook-shaped constellation Scorpius, which barely scrapes above the horizon on summer evenings. A sight line to Antares starts at Deneb, runs through the Northern Cross's long arm and extends about 80 degrees farther. Antares is distinctly orange, brighter than Altair but fainter than Vega. (The locater arrow from the Northern Cross aims almost directly at Antares. Unavoidable distortion in all-sky maps produces an offset and a slight curve in the Vega-Arcturus line, which is actually straight in the real sky.)

In Greek, Antares means rival of Mars, and the name fits. When the reddish planet rides the ecliptic in this region, the two look almost identical. Antares' red-orange color originates with the star. Only half the average temperature of our Sun but with a diameter about 500 times as great, Antares is a rare red supergiant. If Antares were to replace the Sun, it would easily enclose the Earth's orbit. If it were at Vega's distance of 25 light-

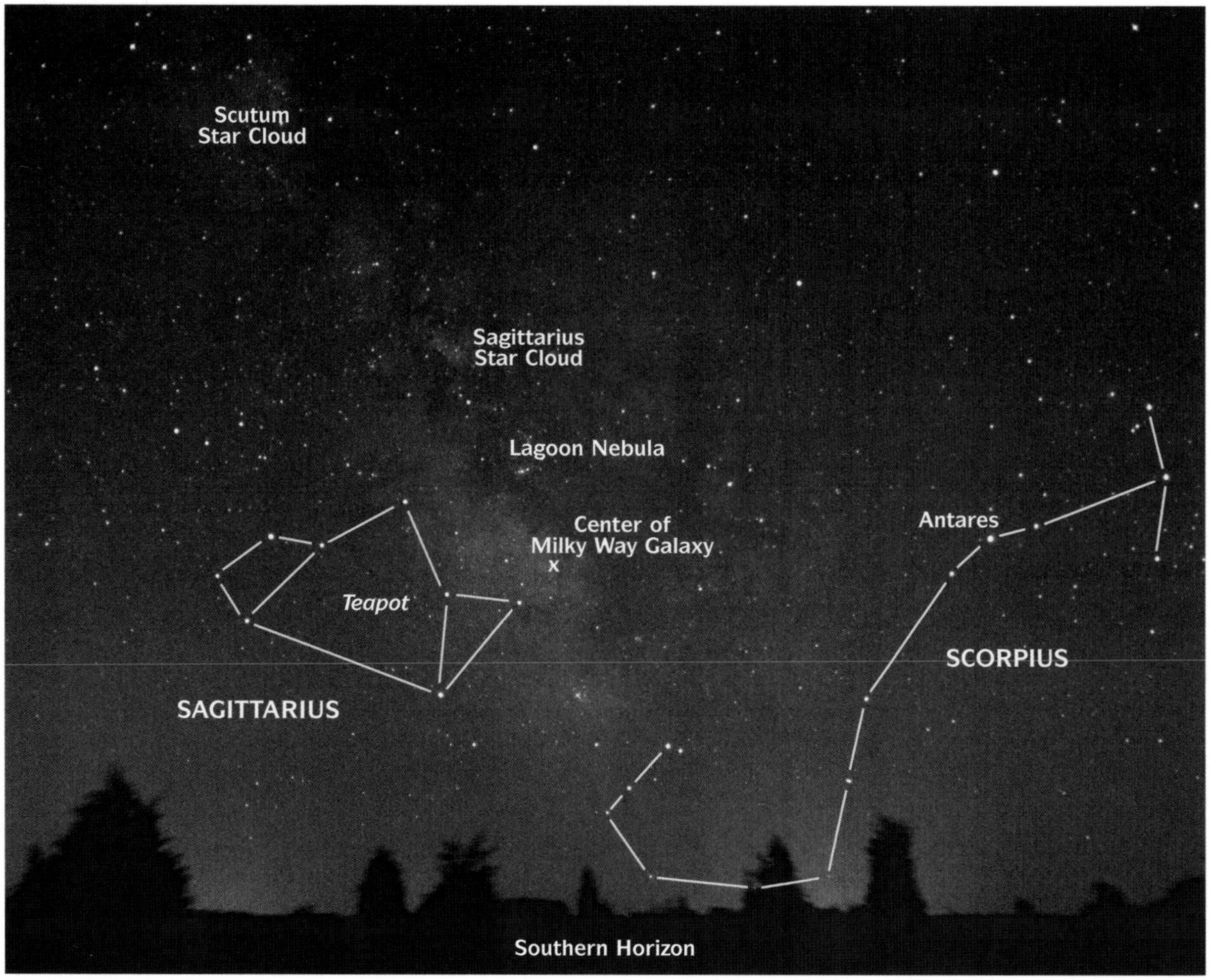

The center of the Milky Way Galaxy is located just off the spout of the Sagittarius teapot. Dozens of galactic delights in this region of the summer sky are visible to the recreational astronomer equipped with either binoculars or a telescope. This photograph was taken from the author's backyard at 44 degrees north latitude, where the tail of Scorpius just skims the horizon. Observers located farther south see Sagittarius and Scorpius higher in the sky, a real advantage for exploring this rich celestial territory.

Urban Myths of Stargazing

An incident that emphasizes just how far some city dwellers are removed from real stars occurred in the hours following a major Los Angeles-area earthquake in 1994. The 4 a.m. quake, centered in Northridge, California, had prompted almost everybody who felt it to rush outdoors for safety and to inspect the damage. But the trembling landscape had also knocked out power over a wide area.

Standing outside in total darkness for the first time in memory, hundreds of thousands of people saw a sky untarnished by city lights. That night and over the next few weeks, emergency organizations as well as observatories and radio stations in the L.A. area received hundreds of calls from people wondering whether the sudden brightening of the stars and the appearance of a "silver cloud" (the Milky Way) had caused the quake. Such a reaction can come only from people who have never seen the night sky away from city lights.

According to Ed Krupp, director of the Griffith Observatory in Los Angeles, many of the anxious callers were reluctant to believe that what they had seen while the power was off was the normal appearance of the real night sky. Krupp, an expert in sky mythology and constellation lore, says that a new mythology has appeared over the past 25 years, a period which coincides with a massive increase in the quantity and brightness of outdoor lighting fixtures.

"Since so many of us never see a non-light-polluted night sky from one year to the next," he explains, "a mythology about what people *think* a true star-filled sky looks like has emerged."

This has spawned what I call urban star myths—generally accepted "facts" about the appearance of the night sky—that can be proved false by just looking at a starry night sky. But, of course, that's the problem.

An example of an urban star myth?

How about this: "The North Star is the brightest star in the sky." Although this is a commonly accepted "fact," the reality is that Polaris, the North Star, is always outshone by 15 to 25 brighter stars, depending on the time of year.

Nighttime lighting is now so pervasive that a pristine view of the Milky Way, left, is a rare sight for many people. The glow from large urban areas is actually quite astonishing, as the photograph above demonstrates. It was taken from the Mojave Desert, 155 kilometers from Los Angeles. Yet the horizon glow from the great metropolis is blatantly obvious. Comet Hyakutake is the small, elongated object just above center.

The Light-Pollution Factor

Children growing up in the final decades of the 20th century or the early years of the 21st century represent the first generation in the history of civilization to live in a world where the stars are almost certain to be the *last* thing noticed at night instead of the first. The change has been swift. Many people 55 or older still clearly remember when the splendor of a dark night sky dusted by the pale glow of the Milky Way was as close as the back door, regardless of where they lived.

Today, outdoor lighting is a fact of life, a ubiquitous background that is as much a part of urban existence as paved roads and shopping malls. Yet all it takes is one attempt to identify a few stars, and your awareness of night lighting abruptly sharpens. From within a city (large or small), you will see the sky as yellowish gray, not black. Outdoor lighting illuminates the air as well as the ground.

To demonstrate the extreme difference between city and country starscapes, I took the two pictures at left just a few days apart using the same camera, lens, film and exposure (25 seconds). The identical sector of the sky is shown. The sky was moonless and exceptionally clear in both cases; the only difference was location. One photograph was taken in rural Ontario, well away from metropolitan areas. The other is a view from my mother-in-law's condominium, which faces Toronto (population four million) from near the edge of the city. (The motion of the planet Mars is apparent between photographs.)

Apart from general urban sky glow, almost every stargazer comes to hate the one or two local lights that seem to shine directly in the eyes. The offending luminaire is usually a streetlight, but porch lamps and dusk-to-dawn "security" lighting are often the source as well. Since most people spend as little time as possible standing outside at night, they never notice that most outdoor lighting produces this glare. It arises from fixtures that are poorly designed or poorly installed, pumping light in all directions instead of limiting it to the intended target. Rarely is there a need to direct *any* nighttime illumination horizontally or higher. It is pure wasted energy—light pollution.

Light pollution is not trivial. Estimates suggest that the energy wasted in North America amounts to approximately one billion dollars a year. That's money spent to generate electricity for light which never touches the ground but, instead, uselessly illuminates the night sky and robs us of a clearer view of nature.

What can you do? Invite your friends and neighbors to look through your telescope. Once they turn their eyes skyward, the problem of glaring lights often becomes self-evident. Probably the most meaningful impact you can have is to set an example by switching your own outdoor lights off most of the time or by converting them to fixtures with an infrared motion detector so that they will turn on only when something moves in the vicinity. Infrared systems save energy and are much more likely to deter someone nasty than a steady glow. For your own use, they come on when you step outside or arrive home and can be overridden by the normal light switch when you are outside observing.

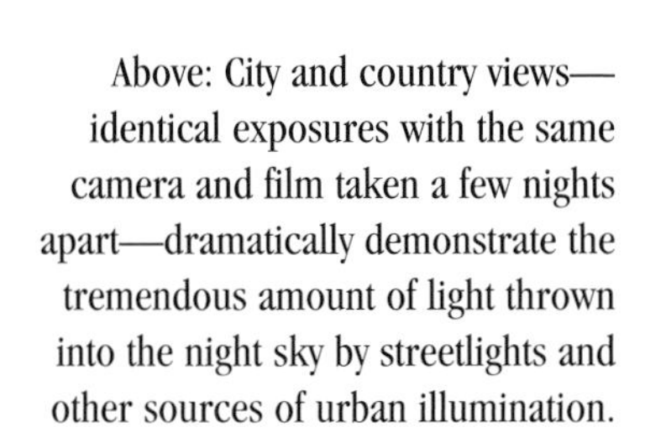

Above: City and country views—identical exposures with the same camera and film taken a few nights apart—dramatically demonstrate the tremendous amount of light thrown into the night sky by streetlights and other sources of urban illumination.

Right: Nature's night light, the full Moon vies for dominance with thousands of streetlamps in today's urban environments.

years, instead of 600, it would be magnitude –6, by far the brightest object in the sky after the Moon.

Also riding low in southern summer skies is the teapot-shaped constellation Sagittarius. The teapot's spout is to the right, its handle to the left. Identify this constellation by running a sight line from Deneb through the Summer Triangle, just to the right of Altair, then extend it to near the south horizon.

The Summer Triangle stars demonstrate the amazing variety of suns that populate the Milky Way Galaxy. Altair, second brightest of the three, is the nearest, some 17 light-years away. Altair is basically like our Sun but about 10 times brighter. To the unaided eye and in binoculars, Altair appears white, similar to the color of light radiated by the Sun.

Vega, the brightest member of the Summer Triangle, is 25 light-years away. Astronomers estimate that Vega is 58 times brighter than the Sun. This is partly because Vega's surface temperature is twice as hot as the Sun's. Higher temperature means that more energy is released per unit area and that the star shines bluish white.

The difference between Altair and Vega should be noticeable to the unaided eye—Altair's white contrasting with Vega's bluish white. For a more obvious color contrast, compare Arcturus (yellow) and Antares (orange) to Vega. The hottest stars are blue, the coolest reddish orange. Deneb, the dimmest star in the Triangle, is the same bluish white color as Vega but, in reality, is by far the most luminous of the three.

Deneb is so far from the Sun—about 1,600 light-years—that astronomers are uncertain how bright it is, but their best estimates place it at 60,000 times the power output of our Sun, making it one of the intrinsically brightest stars in the entire galaxy. Such an amazing superstar so dwarfs the Sun that if Deneb were to be in its place, Earth could be as far away as Pluto and still receive five times more heat and light than it now does.

The guideposts just described provide a foundation for seeking the less prominent stars and constellations of summer. This is the key to the star-and-constellation identification technique: Always start with the brightest, most obvious stars, identifying and linking them with locater arrows, then fill in the details. You are then prepared to branch out to the less conspicuous constellations and star groups.

One of the great pleasures of amateur astronomy is scanning the summer Milky Way through binoculars. Arcing across the summer sky from northeast to southwest, the pale misty band is at its best around midnight in July and earlier in the evening in August and September. What appears as a cloudlike ribbon to the unaided eye is transformed by binoculars into a glittering river of thousands and thousands of stars. Use a padded reclining lawn chair or an inflated child's dinghy, lie back and *slowly* sweep the Milky Way's glorious star fields with your binoculars. I guarantee that you'll be amazed the first time you try it.

The Milky Way appears misty to the unaided eye because of the eye's inability to resolve it into its individual stars. Actually, we are looking edge-on into the densest part of the wheel-shaped Milky Way Galaxy. The center of the galaxy is very close to the tip of the spout on the Sagittarius teapot but 300 times farther away. The Milky Way is particularly rich in this vicinity, although it would be several thousand times brighter if dense clouds of gas and dust did not obstruct our view to the nucleus.

After you have done some general binocular exploration, noting the rifts and clouds of stars in the Milky Way as well as the comparative lack of stars elsewhere in the sky, there are some specific binocular sights worth pursuing. In the constellation Lyra, for example, binoculars will reveal the star closest to Vega, in the direction of Deneb, as a beautiful twin star. (In fact, people with sharp eyesight will see these two stars without any optical aid.) This is Epsilon (ε) Lyrae, a fascinating star system that is detailed on Chart 10 in Chapter 6.

Another region that should be on any binocular sky tour is the rich zone between the end of the fishhook of Scorpius and the spout of the Sagittarius teapot. Especially beautiful are two clusters of dozens of stars, like swarms of fireflies in the night. (Unfortunately, this star-studded region is too close to the southern horizon for a decent view from north of 48 degrees latitude.)

Binoculars also enhance the view when one is observing from the city. Just as in the country, where they can show stars fainter than those visible to the unaided eye, binoculars in the city reveal stars that may be suppressed below naked-eye level by smog and artificial lights. The stars of Lyra, for instance, show up clearly in binoculars from urban locations when only Vega is visible to the unaided eye.

Comparatively few people live where the Milky Way is visible from just outside the door, but that hasn't blunted interest in stargazing and astronomy. More enthusiasts than ever before are seeking out sites that offer stargazing opportunities.

Summer

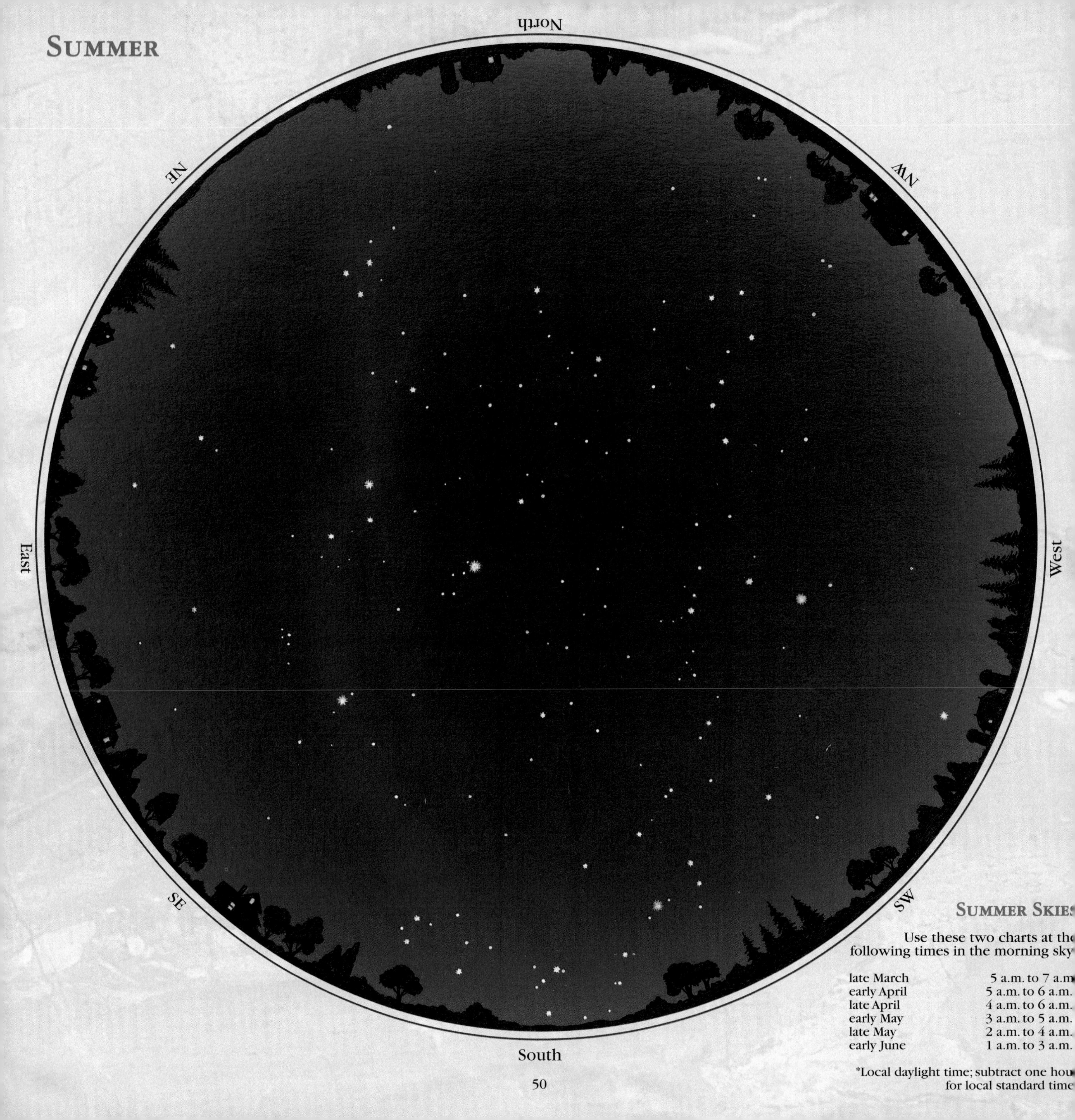

Summer Skies

Use these two charts at the following times in the morning sky*

late March	5 a.m. to 7 a.m.
early April	5 a.m. to 6 a.m.
late April	4 a.m. to 6 a.m.
early May	3 a.m. to 5 a.m.
late May	2 a.m. to 4 a.m.
early June	1 a.m. to 3 a.m.

*Local daylight time; subtract one hour for local standard time

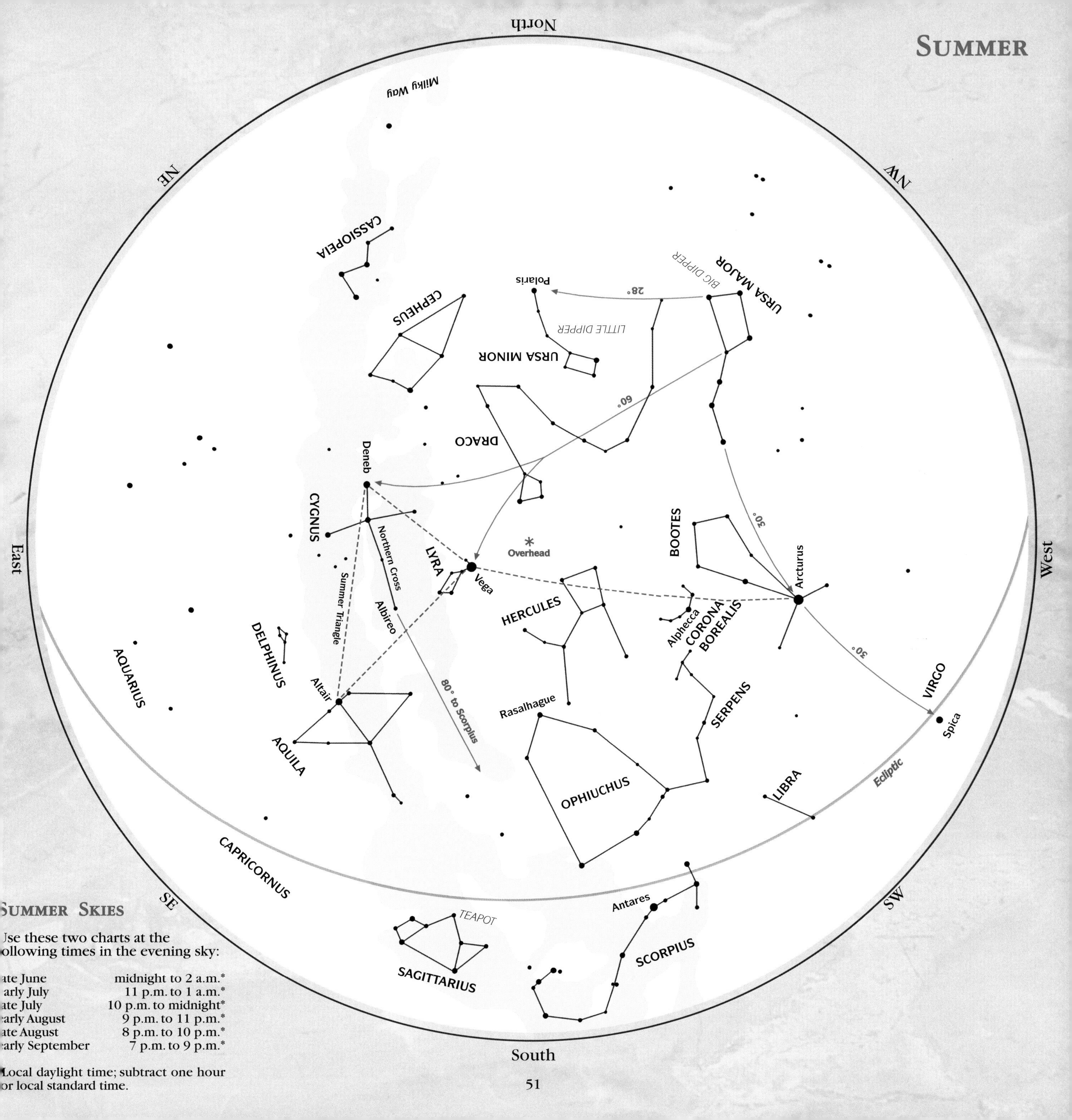

ummer Skies

Jse these two charts at the
ollowing times in the evening sky:

ate June	midnight to 2 a.m.*
arly July	11 p.m. to 1 a.m.*
ate July	10 p.m. to midnight*
arly August	9 p.m. to 11 p.m.*
ate August	8 p.m. to 10 p.m.*
arly September	7 p.m. to 9 p.m.*

Local daylight time; subtract one hour
or local standard time.

The Autumn Sky

Autumn's long evenings and generally comfortable weather for observing combine to produce perfect conditions for backyard skygazing. In June and early July, it is often not dark enough for convenient skygazing until 10 p.m., whereas in October, two extra hours of evening darkness permit leisurely investigation of the night sky.

The autumn sky contains fewer bright stars and distinctive stellar configurations than can be seen during the other three seasons. However, compensation is afforded by more than a dozen second-magnitude stars, which form easily recognized star groups harboring some of the sky's greatest wonders.

The Big Dipper scrapes low toward the northern horizon during autumn evenings, so a dark sky and an unobstructed horizon in that direction are prerequisites for using the Big Dipper's locater-arrow system. The key autumn locater arrow is the one that emerges from the third star in the Dipper's handle and passes through Polaris to Cassiopeia, near overhead, a total of about 55 degrees. If the sight line from the Big Dipper is obscured by trees or lights in the north, scan the overhead region for the distinctive W-shape of Cassiopeia. The constellation is about 15 degrees wide, with each arm of the W three to four degrees long. Cassiopeia, the mythological queen, governs autumn-sky identification. No fewer than four locater arrows emanate from this small configuration, the most important arrow extending south about 35 degrees to the center of the "square" of Pegasus.

The square is fairly large —its sides range from 14 to 17 degrees in length and are marked by four second-magnitude stars. When the observer faces south, the right side of the square forms a sight line south to Fomalhaut, a first-magnitude star near the horizon. The left side of the square similarly aims down to the second-magnitude star Diphda, in the gigantic but dim constellation Cetus. The nearly blank southeastern quadrant of the sky is what I call the Cetus void. Of all the areas in the sky that do not have any first- or second-magnitude stars, this is the largest. The top margin of the Cetus void contains the tiny zodiac constellation Aries, whose brightest star, Hamal, is at the end of an isosceles triangle connected to the eastern side of the square.

The star in the square closest to Cassiopeia does not officially belong to Pegasus. It is part of the constellation Andromeda, whose stars angle up to the northeast. Andromeda's claim to fame is that it contains the Andromeda Galaxy, the most remote object visible to the unaided eye. The triangle of stars forming the half of the W closest to Pegasus acts as an arrowhead pointing southward 15 degrees to the Andromeda Galaxy.

The Andromeda Galaxy is a faint fourth-magnitude smudge, but moonless dark skies are needed to see it. It is almost precisely overhead on November evenings, appearing like an oval erasure mark on the blackboard of the sky. At times, it seems not to be there at all, tantalizingly at the threshold of vision. This fragile, hazy patch is an enormous swarm of suns so remote that the combined energy of its 500 billion stars barely produces a detectable image in the eye.

Queen Cassiopeia's king is Cepheus, whose third- and fourth-magnitude stars form the shape of a kindergartner's drawing of a house. Between Cassiopeia and the adjacent constellation Perseus is a neat twin star cluster called the Double Cluster. Slightly easier to see with the unaided eye than the Andromeda Galaxy, it is a distinct hazy patch. The dual nature of the Double Cluster is clearly evident in binoculars, which just reveal the brighter individual stars of both clusters.

The Double Cluster is 7,000 light-years away, in the Milky Way spiral arm beyond our own. Cassiopeia and Perseus are in almost the opposite direction from the center of the galaxy, but the Milky Way is still rich and impressive in this region.

Blue-white Vega, the brightest star in the Summer Triangle, is the dominant star in the small but distinctive constellation Lyra the harp.

The Winter Sky

It is often said that the stars shine more brightly on crisp, clear winter nights than at any other time. Although the stars may look brighter, actual measurements prove that there is no difference in clarity between the best skies in winter and those at other times of the year. The real difference in the winter sky which accounts for the perception that the stars are shining more brightly is that there are more bright stars. Thus it's the number of bright stars visible that makes the difference rather than any seasonal attributes of winter air.

Lots of bright stars should mean more lavishly appointed star groups. Sure enough, the sky's most impressive constellation,

The Ecliptic & the Zodiac

Although we have dispensed with grids and celestial coordinates on our charts, one vital line remains: the ecliptic, the boulevard of the celestial wanderers—the Sun, the Moon and the planets. In an oversimplified sense, our solar system is like the surface of a wide, circular racetrack, and the planets are the racecars. Sometimes Venus overtakes Earth, sometimes Earth passes Mars; but all the action happens in the same horizontal plane. Thus the planets, the Sun and the Moon are always seen in a restricted band in the sky corresponding to this plane.

The solar system's flatness can be traced back to its origin, about five billion years ago, in a vast pizza-shaped cloud of cosmic dust and gas. Today, the solar system retains the thin disk shape of its birth cloud. It is so flat that the Moon and the planets are seldom more than a few degrees away from the ecliptic.

The stars that form the backdrop for the ecliptic are known as the zodiac, a band of 12 constellations made familiar by the ubiquitous newspaper horoscopes. The word zodiac derives from the Greek *zodiakos kyklos*, meaning circle of animals. However, there are only 7½ animal signs among the 12 constellations: Aries the ram, Taurus the bull, Cancer the crab, Leo the lion, Scorpius the scorpion, Capricornus the goat and Pisces the fish; the half-and-half constellation is Sagittarius the archer—half man, half horse. Then there are human figures: Aquarius the water carrier, Gemini the twins and Virgo the maiden. Of course, these are animals too, but Libra the balance does not fit into even the loosest animal definition. Libra seems to be a later revision intended to distinguish this region of the sky from its former designation, the claws of the scorpion.

Some of the zodiac constellations are faint, despite their important location on the pathway of the planets. To coordinate with the planet tables in Chapter 7, every zodiac constellation is named on the all-sky charts, although some of them are purposely not shown to avoid giving the impression that they are as distinctive as the other star groups. However, all zodiac constellations are shown in the set of more detailed charts in Chapter 6.

The zodiac constellations can be traced back to at least 3300 B.C. Drawings on Mesopotamian artifacts from that time depict Leo and Taurus in combat. Clearly, the Mesopotamian artists were portraying the constellations, because the drawings are adorned with star symbols. The late Michael Ovenden of the University of British Columbia, an expert on the origin of the constellations, deduced that most of the zodiac was designed by about 2600 B.C. Far from being invented by shepherds and nomads as an amusing pastime, the constellations of the zodiac were carefully selected to define and describe, in a useful way, the positions of the Sun, the Moon and the planets.

Having one practical celestial mapping system would have been of vital importance to ancient sailors. The most astute sailors of 4,500 years ago were the Minoans of Crete, who may have convinced other cultures of the value of a single standardized code of sky geography. The Greeks and, eventually, the Romans refined the system into the form still in use today, complete with its mythology and names.

The original division of the zodiac into 12 constellations could have emerged when early skywatchers noticed that Jupiter requires 12 years to complete its trip around the ecliptic, spending one year in each zodiac constellation. Jupiter is the brightest planet seen throughout the night and must therefore have been an object of great interest. (Venus is brighter but is visible for only a few hours before sunrise or after sundown.) Far back in the mists of antiquity, the number 12 became a powerful symbol: 12 apostles, 12 biblical patriarchs, 12 jurors and, of course, 12 months in a year. This final division is most likely a product of the lunar orbit: the Moon sweeps around Earth 12 times a year, with 12 days left over.

The specific groups of stars associated with each zodiac constellation vary greatly in size. Virgo, the largest, is three times the size of Aries, the smallest. Astronomers are unconcerned about the discrepancies that arise from the stellar configurations developed so long ago. But ancient astrologers created the *signs of the zodiac*, each 30 degrees wide, to overcome these inequalities. Modern-day astrologers refer to the signs of the zodiac in horoscopes and elsewhere when they mention Taurus, Gemini and the others. They are *not* referring to the constellations. Two thousand years ago, the signs and the constellations were approximately aligned, but a slow oscillation of the Earth's axial orientation has shifted the constellations west relative to the seasons, so today, the signs do not coincide with the constellations. When an astrologer says that Mars is in Gemini, for example, the red planet is really among the stars of Taurus. Astrology is based on a system that no longer reflects the true nature of the sky.

Autumn

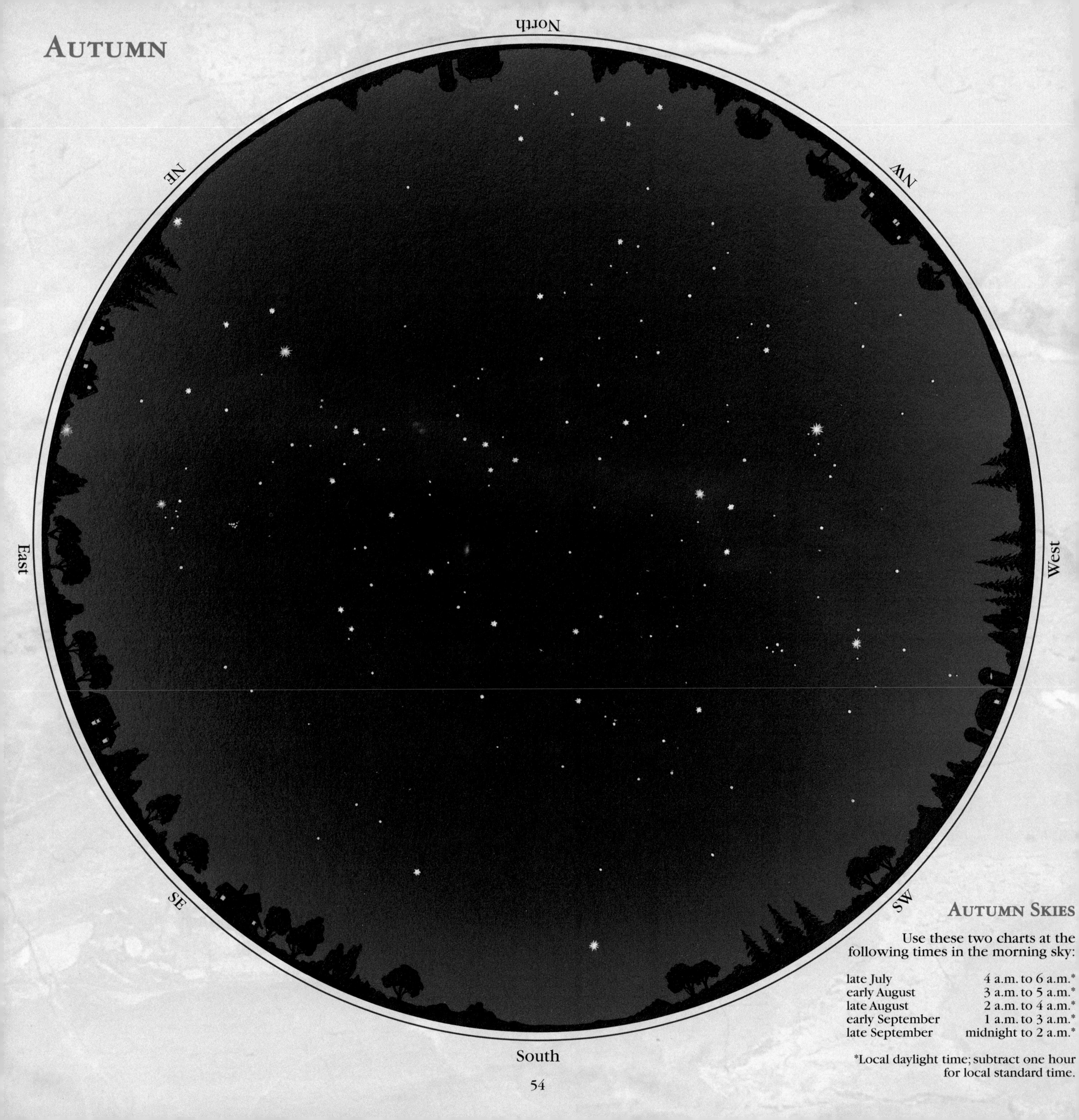

Autumn Skies

Use these two charts at the following times in the morning sky:

late July	4 a.m. to 6 a.m.*
early August	3 a.m. to 5 a.m.*
late August	2 a.m. to 4 a.m.*
early September	1 a.m. to 3 a.m.*
late September	midnight to 2 a.m.*

*Local daylight time; subtract one hour for local standard time.

Autumn

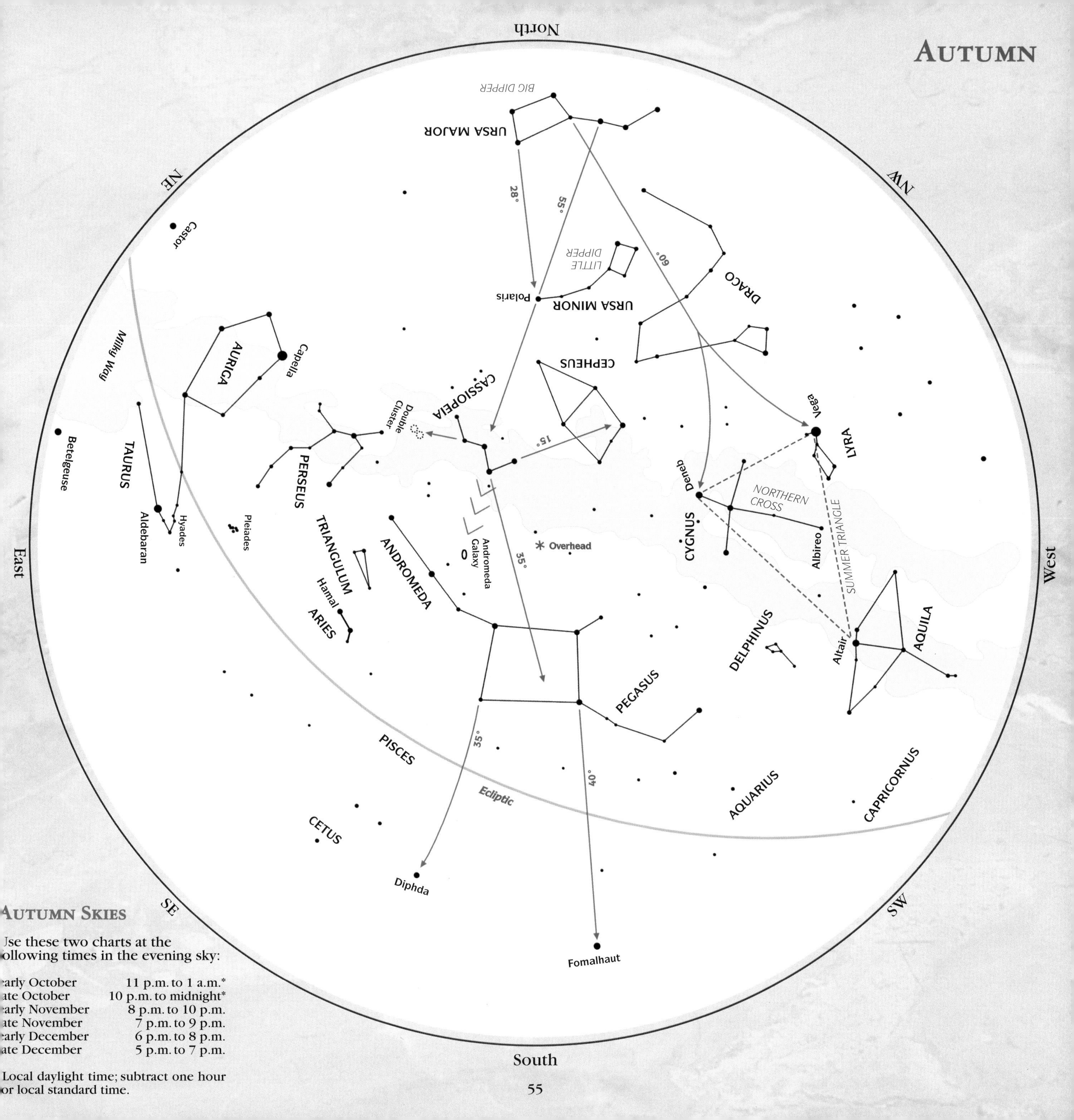

Autumn Skies

Use these two charts at the following times in the evening sky:

Early October	11 p.m. to 1 a.m.*
Late October	10 p.m. to midnight*
Early November	8 p.m. to 10 p.m.
Late November	7 p.m. to 9 p.m.
Early December	6 p.m. to 8 p.m.
Late December	5 p.m. to 7 p.m.

*Local daylight time; subtract one hour for local standard time.

Orion the mighty mythological hunter, lies right in the middle of the winter sky. Orion is the brightest of all the classical star groups, and after the Big Dipper, it ranks as the most distinctive stellar configuration in the heavens. Unlike most constellations, which bear little or no resemblance to their namesakes, the stars of Orion actually look like the outline of a human figure. The nimrod's unmistakable three-star belt is unique. Nowhere else in the sky are three stars of this brightness so close together. Four stars surrounding the belt mark Orion's shoulders and legs.

Rigel, the brightest star in Orion, is one of the most luminous stars known. Shining about 50,000 times more powerfully than the Sun, this hot blue-white stellar beacon is 770 light-years distant. More than a million stars are closer to us than Rigel, but not one of them can match its mighty energy output.

The second brightest of Orion's suns, Betelgeuse, is equally impressive, since it is one of the largest stars known. With an estimated diameter about 800 times greater than the Sun's, Betelgeuse would easily enclose the orbits of Mercury, Venus, Earth and Mars if it were to replace our Sun. It is a member of an exclusive class of rare stars known as red supergiants—obese stellar spheres of low temperature. To the unaided eye, Betelgeuse is distinctly ruddy.

Betelgeuse and the rest of Orion can be found any clear winter evening almost due south, roughly halfway from the horizon to overhead. The belt is about three degrees wide, and the apparent distance from Rigel to Betelgeuse is just under 20 degrees. If the line formed by Orion's belt is extended 20 degrees down to the left, it leads to Sirius, the brightest star in the night sky, at magnitude –1. When the air is turbulent, Sirius appears to twinkle violently, sometimes changing color from white to blue to yellow, offering an almost unending display, like a glittering diamond. When the air is steady, Sirius has a bluish white tinge.

Orion's belt points an equal distance (20 degrees) in the opposite direction to Aldebaran, a first-magnitude star with a yellowish orange tinge. If this pointer from the belt to Aldebaran is continued for another 15 degrees, it leads to a beautiful cluster of stars known as the Pleiades, the seven sisters.

The Pleiades is the brightest and the most distinctive star cluster in the sky. The two-degree-wide stellar jewel box is a wonderful sight in binoculars. Because of its shape, it is often mistaken for the Little Dipper. The real Little Dipper (Ursa Minor) is near the Big Dipper and, in any case, is much less prominent than the Pleiades.

A line running from the middle of Orion's belt straight north through the top of the constellation, between Betelgeuse and Bellatrix, and extending on for some 45 degrees to a point almost exactly overhead comes to Capella, a star second only to Sirius in brightness in the winter sky. From Capella to Sirius, a huge, gently curving arc can be traced that touches three other bright stars: Castor, Pollux and Procyon. These stars also have locater sight lines from Orion.

The diversity of the winter group of bright stars, unmatched in other seasons, provides a good example of the tremendous variation in the celestial zoo of stellar types. Sirius, the brightest, is one of the closest of all stars, only nine light-years away, making it the nearest star visible from Canada and the United States. It is bigger and brighter than the Sun, with about twice the Sun's diameter and 23 times its brilliance. Sirius's exceptional brightness is due to its substantially greater energy output and hotter temperature compared with our Sun.

Capella, Procyon, Rigel and Betelgeuse are all zero-magnitude stars and appear to be almost the same brightness to the unaided eye. And yet this similarity is sheer coincidence. Procyon is only 11 light-years away, Capella is 42, and Betelgeuse is 430, while Rigel is a colossal 770 light-years distant. Furthermore, Betelgeuse is a variable star that oscillates in brightness by almost one magnitude (from 0 to +1). Normally,

Modern techniques of computer processing can coax amazing detail and richness from astrophotos taken on standard color film. The C-shaped nebula to the left of Orion's belt, known as Barnard's Loop, is invisible to human vision, even by telescope.

it is almost as bright as Rigel, but every six years or so, it becomes noticeably dimmer. The last drop was in 1995.

Next in order of brightness in the winter group comes Aldebaran, a first-magnitude star classified as a red giant—a smaller version of Betelgeuse some 65 light-years distant. Aldebaran is 360 times the Sun's brightness and 45 times its diameter and has a distinct orange cast similar to that of Betelgeuse.

Still working through the bright stars in the winter group, we come to Pollux, a first-magnitude star 34 light-years away and 35 times the Sun's luminosity. Its companion, Castor, is officially classified as a second-magnitude star but is almost as bright as Pollux. Castor is 45 light-years away. Because they are so close in apparent brightness, it is often difficult to know which is which. I do it by remembering that Castor is closest to Capella and Pollux is closest to Procyon. The alliteration provides the mnemonic device.

Castor and Pollux are the brightest members of Gemini the twins. Two chains of stars extending toward Orion from these stars form the mythological twin boys. Gemini is famous because it is part of the zodiac, the 12 constellations that happen to be located on the ecliptic. Centuries ago, astrologers (not to be confused with astronomers) first attached importance to these constellations because the Moon, planets and Sun appear to move through them.

A second zodiac constellation in the winter group is Taurus the bull, the bright star Aldebaran marking one eye of the beast. Aldebaran is at the end of one arm of a small V of stars plainly visible in moderately dark skies. This is the Hyades, a star cluster similar to the Pleiades but more spread out and therefore less striking. Binoculars reveal dozens more stars just below the limit of naked-eye visibility.

A unique binocular object, indicated on the chart by "Neb.," is nestled beneath Orion's belt. This is the great Orion Nebula, the most impressive example of thousands of such clouds scattered throughout the galaxy's spiral arms. Look for a small, fuzzy patch in the middle of a short string of stars known as Orion's sword, just below the belt. Binoculars show the nebula's teacup shape—a puff of cosmic gas seemingly frozen in timeless space.

The Milky Way, that misty band of light which is usually associated with summer skies, extends almost entirely across the winter sky from northwest to southeast. This winter sector of the Milky Way is not as bright as its summer counterpart, because we are looking generally toward the outer edge of our galaxy rather than toward its heart.

Some of the best parts of the winter Milky Way are between Orion and Gemini and up through Auriga. Binoculars reveal the true stellar nature of this celestial ribbon of light. Thousands of stars pass through the field of view in places where only a pale haze is seen with the unaided eye.

Swaths of the Milky Way visible in summer, autumn and winter skies act as a reminder that we observe the universe from within a disk of stars. The stellar scene probably was not much different a million or even a billion years ago, because the Sun's path through space keeps it in the vicinity of the spiral arms. As timeless as the vista may be, only in the past few decades has anything close to a comprehensive picture of the cosmos come into focus. The creatures who inhabit one planet of one star of one galaxy among billions now know the extent, if not the ultimate meaning, of the visible universe. Today's backyard astronomers peer out into the void with knowing eyes.

This extreme wide-angle photograph of the winter sky, taken facing south from the author's backyard, shows all the important stars and constellations seen from December through March. Use centrally placed Orion to work your way among the star groups.

Winter

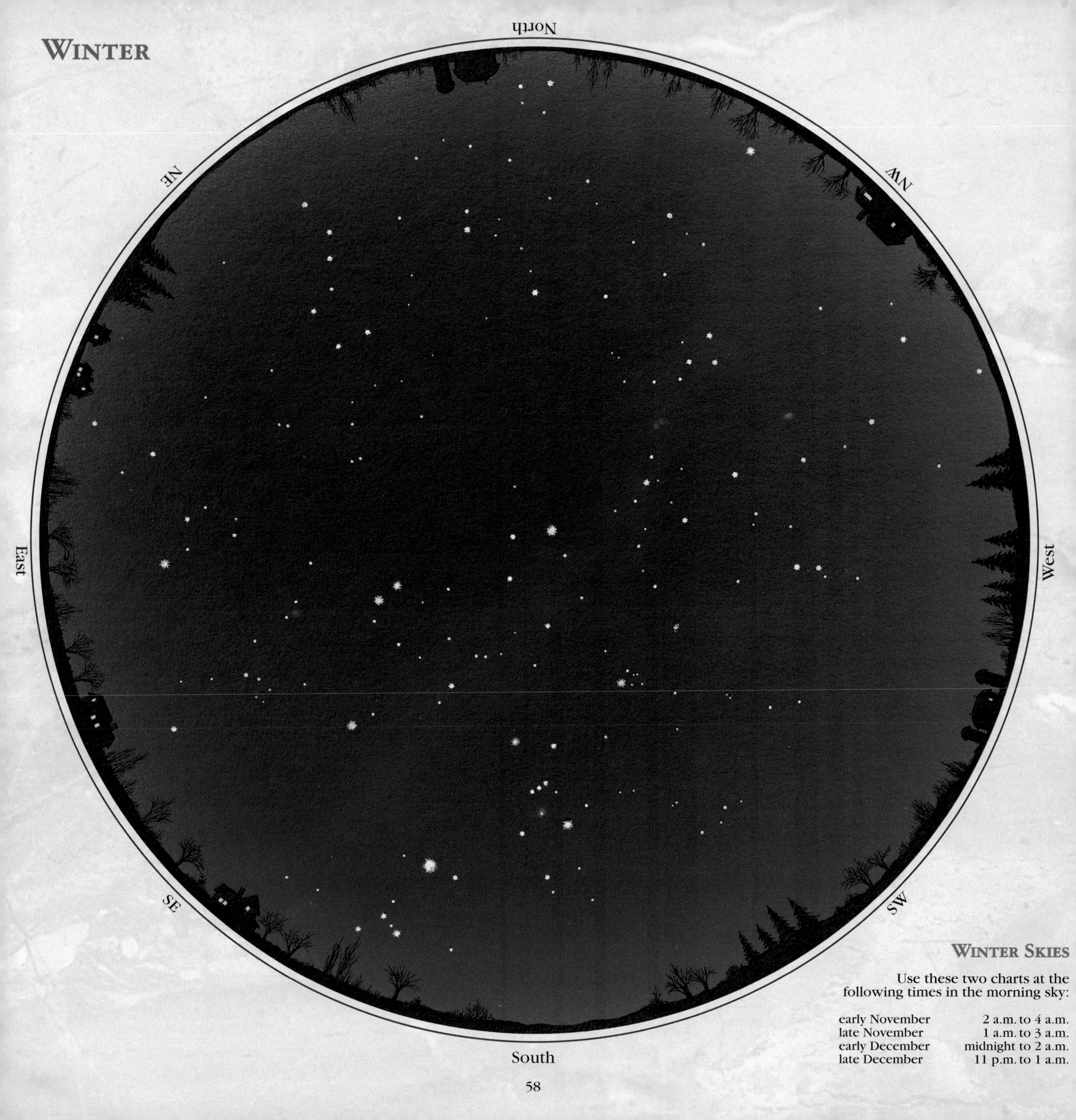

Winter Skies

Use these two charts at the following times in the morning sky:

early November	2 a.m. to 4 a.m.
late November	1 a.m. to 3 a.m.
early December	midnight to 2 a.m.
late December	11 p.m. to 1 a.m.

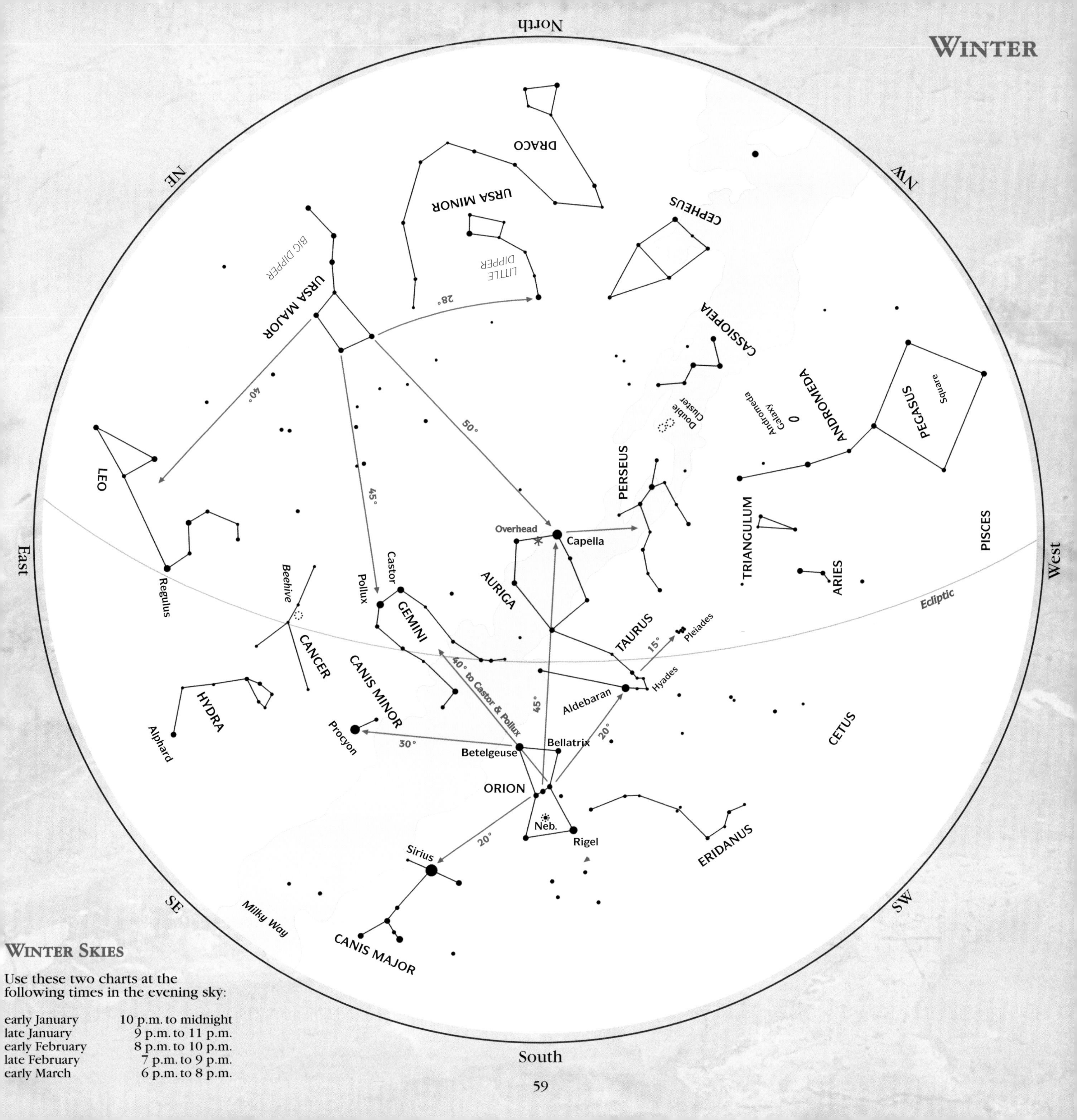

WINTER SKIES

Use these two charts at the following times in the evening sky:

early January	10 p.m. to midnight
late January	9 p.m. to 11 p.m.
early February	8 p.m. to 10 p.m.
late February	7 p.m. to 9 p.m.
early March	6 p.m. to 8 p.m.

CHAPTER FIVE

STARGAZING EQUIPMENT

O, telescope,
instrument of much knowledge,
more precious than any sceptre, is not he who holds thee in his hand
made king and lord of the works of God?

Johannes Kepler

Next to a camera, binoculars are probably the most common optical equipment found in the home. Yet I am always surprised to discover how seldom, if ever, people raise their binoculars to the night sky. Even the most modest binoculars can turn a starry night into a pageant of jeweled velvet as you cruise along the Milky Way. The color of stars also tends to be more intense in binoculars than when viewed with the eyes alone. Binoculars reveal dozens of the Moon's craters along with four of Jupiter's moons. And, if you know where to look, binoculars can transport you millions of light-years to galaxies far beyond our home galaxy, the Milky Way. Not bad for a piece of equipment the size of a hardcover novel!

In fact, binoculars are the perfect training wheels for anyone contemplating the purchase of a telescope. Binoculars actually *are* telescopes—two of them—but more compact than conventional telescopes because the light path is compressed by the use of prisms within the structure of the instrument. The small size and the low magnification of binoculars compared with telescopes place them in a unique category, midway between the telescope and no optical aid at all. I especially like the way I can lie back in a reclining lawn chair or—my favorite—a child-sized inflatable dinghy and leisurely scan the celestial tapestry with these compact mini-telescopes.

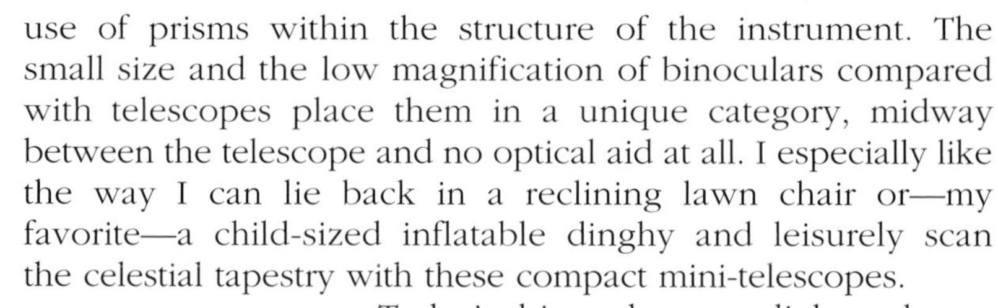

Today's binoculars are lighter, better designed and technically superior to those of the past. If you are using binoculars more than 20 years old, you might want to think about getting a new pair, especially if yours are heavy compared with modern binoculars. Weight is a vital consideration for a hand-held instrument.

The most common binoculars are the 7x35 size, which magnify seven times and have main lenses 35mm in diameter. In effect, they are equivalent to a telescope with a 35mm aperture and a 7-power (7x)

Under perfect conditions, a lone observer begins an evening of personal exploration of the universe.

Today's amateur astronomers can select from a far wider range of quality equipment than they could a generation ago.

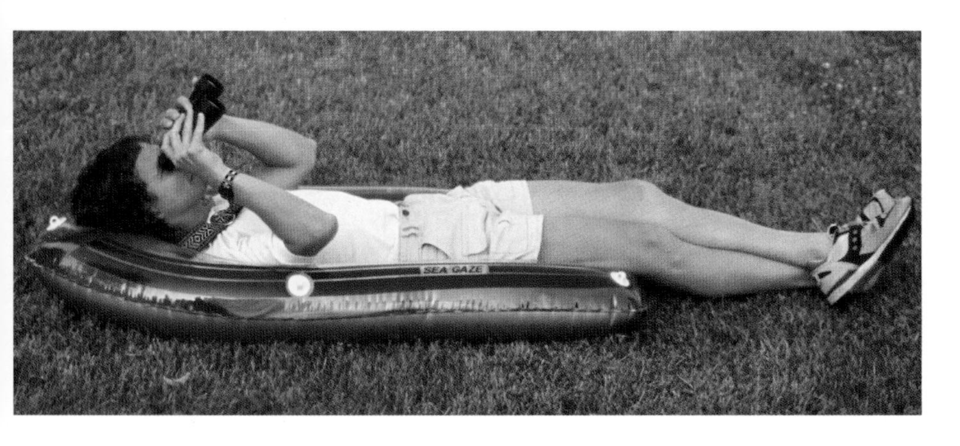

eyepiece. An instrument of this size provides pleasing views of many celestial objects, but given a choice, I prefer slightly larger glasses, such as 7x42, 8x40, 7x50 or 10x50 binoculars. The larger main lenses on these models gather from 130 to 200 percent more light than the 7x35s, often at relatively little additional cost or weight. Good-quality binoculars of this size are able to detect ninth-magnitude stars and reveal all the sights described in this book as suitable for binoculars.

There is always the temptation to try higher-power binoculars ("...if 10-power is good, 20-power should be better"). But there are serious trade-offs. The higher the power, the narrower the field of view. Less is seen at one time when you have a small field of view, and aiming becomes more difficult, especially at night. One of the chief advantages of binoculars is the wide field of view, and it should not be sacrificed. Most binoculars have the field of view engraved on the instrument body, either in degrees or in terms of feet at 1,000 yards or meters at 1,000 meters. We want the scale in degrees so that we can conform to the apparent distance scales introduced in earlier chapters. In recent years, most binocular manufacturers have begun designating the field of view in degrees, but if degrees are not indicated on the binoculars and conversion is necessary, one degree of field is 52 feet at 1,000 yards or 17 meters at 1,000 meters.

The average 7x50 binoculars have a field of view of about seven degrees, much larger than any telescope. Anywhere from five to eight degrees is ideal for astronomy. On the other hand, 20x50 binoculars, with their three-degree field of view, find little favor among astronomy enthusiasts. Aiming becomes a chore, and once the target is acquired, it is impossible to keep the glasses steady, because hand and arm quivers are being magnified 20 times.

The shaking generated when holding binoculars by hand has long been an accepted downside to the overall ease of use of these handy instruments. But the higher the binoculars' magnification, the greater the shakes are amplified. This limits the *useful* magnification of hand-held binoculars to about 10x.

To summarize, then, the ideal binoculars for astronomy have main lenses 40mm to 50mm in diameter, a magnification of no more than 10x and a field of view of five to eight degrees. Other desirable attributes are light weight, fully coated optics (to increase light transmission) and sharp optics.

Selecting Binoculars

First of all, if you already own binoculars with at least 35mm main lenses, they will get you started. But keep in mind that night-sky viewing—imaging bright pinpoints on a black background—is the most stringent test for optics. Binoculars with good optics show stars as tiny specks that do not become elongated or misshapen except near the edge of the field. A few outstanding binoculars (very few!) are sharp right to the edge.

When evaluating binoculars for astronomy, ask yourself the following questions: Are they heavy for their size or otherwise awkward to use? Do they cause eyestrain? Are they difficult to focus? Are objects at the edge of the field distractingly fuzzy, even though the center of the field is in focus?

If you answered yes to any of these questions, you should reject those binoculars. Not only do top-quality binoculars produce tiny pinpoint star images, but they have antireflection lens coatings to increase light transmission and reduce internal reflec-

With magnifications and fields of view midway between those of telescopes and no optical aid at all, binoculars are essential stargazing tools. One of the best binocular accessories is a child's inflatable dinghy, top. Lightweight and highly portable, it provides more comfortable head-and-shoulder support than a lawn chair can, with the additional advantage of pneumatic head-adjustment capability (just apply leg pressure). High-quality binoculars usually have improved lens coatings to reduce light loss by reflection. The bottom pair of binoculars at left has the superior coatings. Right: A threaded tripod adapter hole, an important feature for astronomy, is usually concealed behind a small screw-on cap.

tions that cause ghost images. The lenses and prisms are made to exacting standards and are designed to deliver bright, sharp aberration-suppressed views—and they are expensive. Even so, I have never seen truly perfect binoculars for astronomy (I'm fussy). Some have superb optics but are too heavy to hold; others are jewels in every respect but carry astronomical price tags (up to $2,000).

During tests of dozens of binocular models, I found several with obscure brand names that were excellent performers. Price seems to be a better guide than brand name in this competitive field, so don't reject something on the basis of an unfamiliar name. Expect to pay at least $150 for astronomical-quality binoculars.

Among many fine models I tested, the Bausch & Lomb Legacy 7x50 binoculars are a standout in the $150 to $200 range. Although widely available in Canada, this particular model is not distributed in the United States for some reason. I also like the Celestron Ultima series and the top lines offered by Orion, but there are lots of others. Compare as many models as you can before making your final choice.

For the aficionado who wants the very best, the finest astronomy binocular I have used is the Canon 15x45 IS ($1,500), which utilizes a revolutionary image-stabilization prism that virtually eliminates hand-held shakes and allows effective use of

the relatively high 15 power. Outstanding eyepiece design on the Canon glasses gives a supersharp 4.5-degree field. Two excellent conventional top-of-the-line glasses are the Zeiss 7x42 Dialyt ($900) and the Bausch & Lomb 8x42 Elite ($600).

As a general rule, avoid "zoom" binoculars, because they have little application in astronomy and the optics are often inferior to those of comparably priced fixed-magnification glasses. I also tend to avoid binoculars designated by their manufacturer as "wide angle." While they yield amazing picture-window vistas up to 11 degrees wide, large portions of the outer part of the field in most models remain out of focus due to inherent limitations of wide-angle binocular optics. This is more noticeable in astronomy than in daytime land viewing, and I find the defect objectionable. But the choice is mainly one of personal preference.

When evaluating binoculars for astronomy, one key consideration is the tripod adapter hole—a threaded hole at the front of the central bar.

Whether tripod-mounted or hand-held, binoculars are the preferred instrument to use to become acquainted with the night sky before purchasing a telescope.

An inexpensive L-shaped bracket fits the hole and allows your binoculars to be attached to a camera tripod. Significantly more detail—especially subtle astronomical detail—is visible through binoculars steadied in this way. Any binoculars purchased for stargazing should have the adapter hole. Many do not, so ask about this feature before you buy.

Camera tripods are essential for binoculars over 50mm aperture (with the possible exception of 8x56 models, which are just at the hand-holding limit for most people). Larger glasses are simply too heavy to hold comfortably for more than a few seconds. My 11x80s weigh in at just under 5 pounds, compared with 1.6 pounds for my 10x50s. The brutes also come in 10x70, 16x70, 15x80, 20x80 and several other configurations. When fixed to sturdy camera tripods with L-brackets, these binoculars work well for astronomy and are outstanding for long-distance land viewing. Large binoculars are the transition instruments between hand-held binoculars and telescopes, and I regard them as optional astronomical equipment.

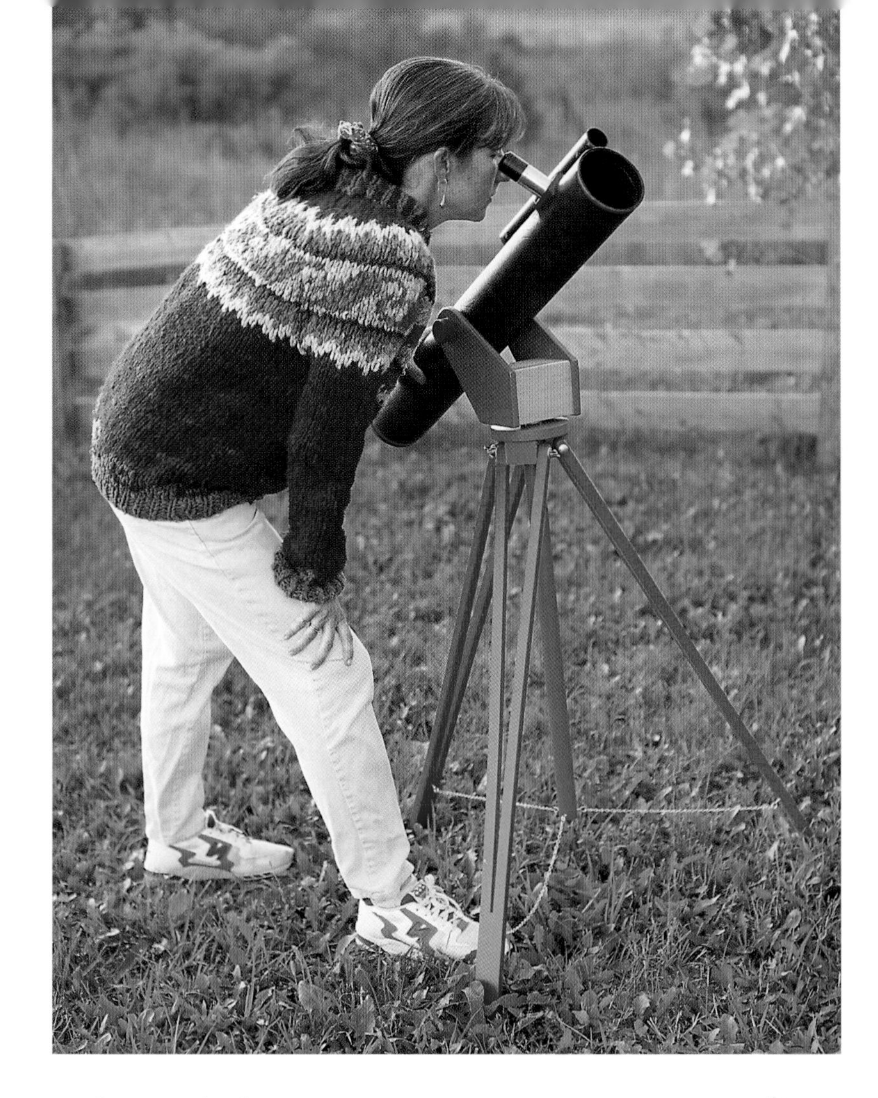

Telescopes

A good pair of binoculars and a shelf full of astronomy reference books and star atlases are the bare essentials for the backyard astronomer. Yet such minimal aids can provide years of stimulating exploration—by mind and eye—into the depths of space. But sooner or later, almost everyone who is captivated by the mystery of the starry night craves a telescope. The problem is, the craving usually starts sooner, rather than later.

More than a million telescopes have been sold in Canada and the United States over the past 25 years. Most of them are inexpensive so-called beginners' telescopes purchased as gifts by parents or spouses or by people with a bubbling enthusiasm for astronomy but little knowledge. My first telescope, bought under exactly those circumstances, was an inexpensive 60mm refractor. (The figure refers to the diameter of the main lens.)

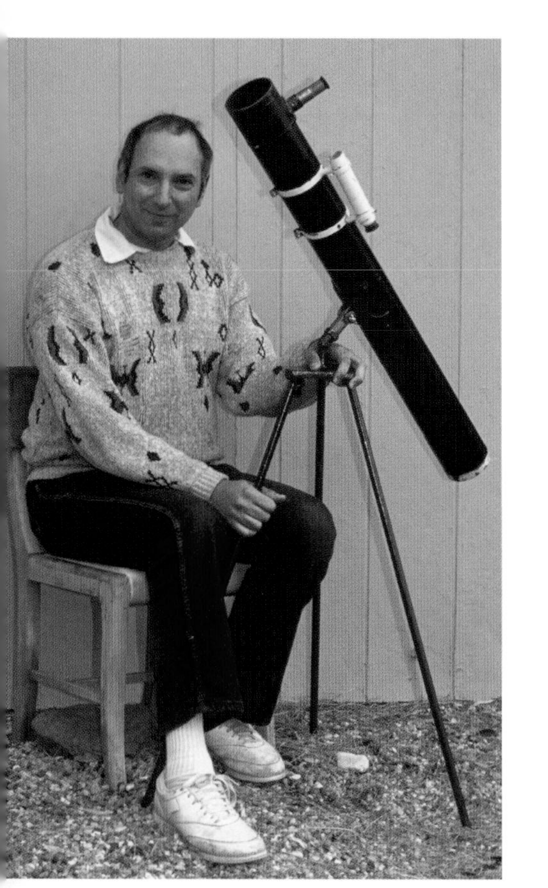

Left: World-famous amateur astronomer David Levy with his first telescope (purchased in the 1950s), a 3-inch Newtonian reflector with decent optics but a rickety mount. The Stargazer Steve unit, top, is a much more stable 1990s version.

That little instrument on its jiggly tripod gave me hundreds of hours of pleasure—and frustration. Pleasure because, for the first time, I could see things that I had read about in astronomy books. Frustration because the telescope would quake and shiver in the slightest breeze, making the object in view resemble the jumping dot of an oscilloscope.

More than anything else, that telescope taught me what to look for in my next telescope. First and foremost, I wanted a rock-steady mount. Second, I wanted larger and better optics that would give brighter, sharper images. However, I also knew from using the small telescope not to pay any attention to the manufacturers' magnification claims. A 60mm refractor, the most common type available then and now, cannot be used to advantage at magnifications exceeding about 120x. It is the same principle as a car speedometer which shows a maximum speed that is 50 miles per hour faster than the car can possibly go.

But I learned all that *after* the purchase of my first telescope. And that problem persists today. Inadequate, poorly designed beginners' telescopes are still being sold to well-intentioned but uninformed consumers. Here are a few things to avoid when purchasing a telescope.

The classic novice telescope is most easily identified by its spindly tripod, either full-sized or (worse) tabletop format. The mounts have nice silver knobs and fancy dials, and the accessory boxes are full of gadgets (largely useless). The instruments themselves are most often 50mm or 60mm refractors or 75mm to 115mm (3-to-4.5-inch) reflectors. Priced in the $75 to $300 range, these telescopes are available everywhere: department stores, camera shops, hobby-supply outlets. They may have familiar brand names, but virtually all such telescopes are made in a few giant factories in Asia. I cannot recommend any telescope under $300, no matter whose name is on it. (Notable exception: Stargazer Steve's beginners' telescopes, made in North America.)

It is not necessarily poor optics that are at fault in these telescopes; it is everything else—the eyepieces, mount, tripod, locking screws, slow-motion controls, finderscope and instruction booklet, which all range from poor to abysmal. Some beginners' telescopes are so inadequately designed that they are impossible to use for viewing anything but the Moon.

If these beginners' telescopes are so bad, why are they offered for sale? The answer is simple: where there is a market, there will be a product to fill it. A $200 telescope that comes with a collection of impressive-looking accessories is an attractive gift purchase for someone with good intentions but little knowledge.

Good-quality telescopes that offer a minimum array of appropriate accessories begin around the $400 mark. If you can afford to go significantly over $500, you will greatly reduce your chances of being stuck with something you wish you had not purchased. Plan ahead, and do it right the first time.

There are many advantages to bypassing the classic beginner's telescope, not the least of which is that a better, larger telescope is actually easier to use than the average small department-store telescope. The greater weight and stability of the larger instrument's mount means that

One of the chief defects of so-called beginners' telescopes is inadequate jiggly mounts that make aiming the instrument an exercise in frustration. An economical and effective alternative is the simple and stable Dobsonian mount, above.

The Dobsonian is an example of an altazimuth mount—a telescope mount with simple horizontal and vertical motions only. The other basic category of mount is the equatorial.

Trash-Scope Blues

"Am I doing something wrong, or is it this telescope?" the caller on the telephone asked. "It doesn't seem to focus, and I am never sure what I am pointing at."

I was speaking with another frustrated owner of a $200 department-store telescope, the kind that comes in an attractive package announcing "450-Power Astronomical Telescope." The caller admitted that the impressive complement of accessories included with the instrument as well as the packaging embellished with color photographs of comets and nebulas had been too much to resist. But now, he was wondering whether he had made a mistake.

"I can't seem to keep it steady—everything's a blur," he moaned. "The only eyepiece that shows anything at all is the one marked K20mm. What am I doing wrong?"

"Nothing," I sighed. I had heard it all before, hundreds of times. "It's not you," I assured him, "it's the telescope."

I went on to explain that his department-store 60mm refractor telescope was an example of what many experienced amateur astronomers uncharitably but aptly call "Christmas trash scopes." I told him that his frustration was perfectly normal. The classic trash scope is designed not for ease of use but to appeal to rank beginners and well-meaning gift buyers. Well-known brand names also lure the novice, but they mean nothing, since all telescopes in this class are manufactured in Asia, regardless of the name.

Even expert observers soon become frustrated when trying to operate these instruments, with their jiggly mounts and rickety tripods. They barely function and usually only at their lowest power, typically about 36x with a 20mm eyepiece. Accessories such as Barlow lenses, image erectors, Sun-projection screens, filters and high-power eyepieces (5mm to 12mm) supplied with these telescopes are so cheaply made that they are almost impossible to use and are included merely to give the impression of a fully equipped instrument.

If you have a telescope like this, use it with the 15mm to 25mm low-power eyepiece and the right-angle prism diagonal for introductory views of the Moon, Saturn's rings, Jupiter's moons and a few double stars. Don't expect much more. And if you do not already own a telescope like this, consider yourself lucky to have read this first.

Frequently Asked Questions About Telescopes

Which type of telescope is best: refractor, Newtonian reflector or Schmidt-Cassegrain?

Much of this chapter is devoted to comparing the positive and negative aspects of various telescope designs. Please read it carefully before investing in a new telescope. The short answer is that each type performs admirably—if it is manufactured to sufficiently high standards. Fortunately, most commercial telescopes meet this criterion (exception: the "trash scopes" referred to on page 65). The telescope market has become very competitive in recent years, and the familiar adage "you get what you pay for" is a useful starting point.

Can a nature spotting scope substitute for an astronomical telescope?

Not really. These instruments are made for daytime observing and, considering their cost, are a poor substitute for an astronomical telescope of the same price. For one thing, they are usually configured for straight-through viewing —not very convenient for looking up. However, if a nature scope is all you have, it's certainly better than no telescope at all.

Can I see where the astronauts landed on the Moon?

Yes, the corner of the Sea of Tranquillity, where the first humans landed on the Moon in 1969, is easy to spot (see Moon maps on pages 140 and 141). But what most people mean by this question is: Can I see the lunar lander? Sorry. No telescope on Earth is powerful enough to do that. The lander is roughly the size of a delivery van—far too small to be detected at 100 times the distance from New York to Los Angeles.

Why are the images upside down?

All astronomical telescopes invert the image. And, to add to the confusion, telescopes with prism or mirror diagonals also flip it left to right. To get things right side up, a prism or extra lenses are required. In astronomy, though, we want as "pure" an image as possible, using the optimum number of optical elements. An erect image counts for land viewing but is largely irrelevant in astronomy. This is always disorienting for the beginner, but with practice, you'll get used to it.

Can I keep warm while observing by looking through a window with a telescope or binoculars?

Trying to avoid cool outside air by pointing a telescope through window glass is almost always futile. Virtually all glass panes introduce distortion to telescopic views. Binoculars are not affected as much, but for critical definition, viewing through windows is a no-no. Opening the window will not work either, at least for telescopes, because the flow of air will generate very poor seeing conditions.

Can I see the rings of Saturn?

Yes. Any telescope with a magnification of 30x or more will show this magnificent adornment of the sixth planet. Saturn is unforgettable in a good-quality backyard astronomer's telescope .

How do I align the finderscope?

During daylight, aim the telescope with a low-power eyepiece at a distant chimney or antenna, then use the adjusting screws to set the finderscope's cross hairs so that they point at exactly the same spot the main telescope does. Tighten the finderscope adjusting screws. That's all there is to it.

How often should I clean the optics?

Too much cleaning is worse than too little. There is always an opportunity to scratch a lens or mirror during casual cleaning. Dust accumulation on the lenses and mirrors is inevitable and will not noticeably affect performance if you keep it to a minimum by covering the optics when not in use. If you must clean, first remove loose dust with a lens blower, then use a cotton ball dampened with distilled water to remove dust and stains. Never rub hard. Gently dry with a clean, lint-free lens cloth. Telescope mirrors are especially delicate and should be cleaned as rarely as possible. For eyepieces, use a Q-tip dipped in camera-lens cleaning fluid. Follow with a dry Q-tip. If in doubt, leave it.

Why don't stars look like tiny disks?

Alpha Centauri, the nearest star beyond our solar system, is the same size as the Sun but 250,000 times farther away. That makes it much too small to appear as a disk in even the largest telescopes. Some stars (Betelgeuse, for example) are hundreds of times larger than the Sun but are also far more distant than Alpha Centauri.

What are the numbered dials on the mount for?

These dials are the right-ascension and declination circles. They can be used to aim the telescope at a celestial target. In practice, though, the dials on beginner-level telescopes seldom have the accuracy required to find anything. They are there mainly to give the instrument a "scientific" appearance.

Apparent Sky Motion

once the telescope is aimed at an object, it stays where it is pointed. But that's not the whole story. All celestial objects move, because we are observing them from a planet that rotates once a day. The motion is visible as soon as you look into the eyepiece—the target object methodically marches through the field of view. The higher the magnification used on the telescope, the faster the target moves.

Tracking a celestial object is accomplished in one of three ways, depending on the type of telescope mount: (1) pushing the tube; (2) turning the knobs of slow-motion controls; (3) letting the telescope mount's motor drives track the object across the sky. All three methods work well if the telescope is suitably designed. For instance, pushing the tube sounds too crude to be an acceptable technique, yet this is exactly how telescopes on Dobsonian mounts are designed to work. Altazimuth mounts, commonly supplied with 70mm to 90mm refractors, often have slow-motion control knobs that smoothly adjust the up-down and left-right motions.

Equatorial mounts equipped with motor drives are geared to compensate for the Earth's rotation. Once the polar axis of the equatorial mount is aligned toward the north celestial pole, near the star Polaris, the celestial target can be observed without having to make constant adjustments to keep it centered in the field of view. But there are trade-offs with each tracking technique. For example, equatorial mounts are heavier and more expensive, especially when equipped with motor drives. As we proceed through this chapter, I will outline the options available to the beginner and will offer some recommendations.

Finderscopes

Finding celestial objects is usually easier with bigger telescopes because they are often equipped with adequate finderscopes—miniature telescopes mounted parallel to the main tube that allow easy alignment on the target object. Once centered in the cross hairs of the finderscope, a celestial body is automatically centered in the main telescope.

The finderscope has a much wider view of the sky than the main instrument has, making it easier to zero in on the desired object. Finderscopes on less expensive telescopes usually have inadequate optics and mounting brackets barely stronger than paper clips, both of which make it difficult to align the finderscope with the main telescope. Something as fundamental as locating a desired sky object then becomes a frustrating experience. "The larger, the better" is the maxim for finderscopes. The 5x24 size is inadequate; 6x30 is barely acceptable; 8x40 and 8x50 are far superior. An alternative to finderscope upgrading is the Telrad ($50), an illuminated sighting device easily added to any telescope.

Don't make the same mistake I did by purchasing an inferior beginner's telescope. Be patient, learn your way around the sky with binoculars first, and save your money for a larger good-quality telescope.

Telescope Types

There are three main types of telescope optical systems used in backyard astronomy: refractors, Newto-

Many beginners' telescopes are fitted with equatorial mounts designed to track celestial objects. The observer sets up the mount so that its polar axis is aimed at the celestial north pole, above. High-precision alignment is required only for astrophotography. To compensate for the motion caused by the Earth's rotation, the observer just turns a single knob—the polar-axis slow-motion control. If the telescope is fitted with a polar-axis motor drive, the tracking is automatic. The numbered dials seen on the mounts of many introductory telescopes are mostly intended to impress the novice. They seldom have the required accuracy to point the telescope effectively at a celestial target.

nian reflectors and Schmidt-Cassegrains. The Maksutov-Cassegrain, a relative of the Schmidt-Cassegrain, has also become fairly common in recent years.

Refractors of all sizes are high-tech spyglasses that have a main lens which concentrates the incoming light to a focus at

the lower end of the telescope tube. There, a magnifying ocular called an eyepiece provides a visual image. Refractors with a main lens between 70mm and 100mm are excellent performers, rugged and reasonably priced.

Refractors are available in two designs: achromatic and apochromatic. Achromatic refractors have a long and distinguished history, having been manufactured for more than 200 years. The main lens is actually two lenses with a thin air space between them. The design works quite well up to an aperture of about 100mm. After that, what is known as residual color (caused by the unequal focusing of different colors by the lens glass) becomes obvious as a purple halo around bright objects and at the edge of the Moon. For this reason, you won't see many achromatic refractors larger than 100mm (4 inches).

To reduce the residual color to insignificant levels, much more expensive glass is required, in either a two-lens or a three-lens configuration. These apochromatic (color-free) refractors, introduced to the amateur-astronomy market during the 1980s, produce exquisitely sharp images and are available in sizes up to 180mm aperture. Apochromatic refractors (also known as apo refractors) are premium instruments at a premium price, but they have become very popular among aficionados of fine optics. They are offered by Astro-Physics, Meade, Celestron, Tele Vue, Vixen and Takahashi. A 100mm apochromatic refractor with equatorial mount sells for $2,000 to $3,000.

The simple and elegant design of the **Newtonian reflector** has kept it in the front ranks of astronomy throughout the 20th century. The Newtonian employs a precision-ground shallow bowl-shaped mirror at the base of an open tube to reflect light and bring it to a focus near the top of the tube. A smaller flat-surface mirror angled at 45 degrees and suspended at the top of the tube reflects the light through a hole in the side, where the focuser is located. With the focuser at the top of the tube, a Newtonian offers a comfortable eyepiece position for overhead viewing, while other types can be at their worst when looking straight up. Because they are easier to manufacture than refractors, Newtonian reflectors can be made in larger sizes at a cost within reach of most backyard astronomers.

As with refractors, the most inexpensive Newtonian reflectors should be avoided. In the 3- and 4-inch size range, they are often equipped with inferior mounts and accessories. Another danger the novice will likely be unaware of, even after purchasing the telescope, is that the mirror system can be jarred out of alignment during transport or with daily use. Realigning the mirrors is not difficult, but the inexperienced observer who sees fuzzy, cometlike images of stars may assume that poor optical quality is to blame. Alignment problems aside, Newtonian telescopes offer outstanding value for the money.

Newtonian reflectors come on two types of mounts—equatorial and Dobsonian. When properly aligned, the equatorial mount will track the stars. Equatorial mounts are functional and manageable for 6-to-8-inch Newtonians, but they become cumbersome for 10-inch telescopes and are downright massive in larger sizes. Transporting one of these behemoths to and from a dark observing site is a major expedition. (The ritual of spending an hour dismantling the telescope at 2 a.m. is enough to discourage all but the most rabid enthusiasts.) This is why Dobsonian mounts invaded the amateur-astronomy scene during the late 1970s, when Newtonians 10 inches

Left: The classic long-tube achromatic refractor is a fine lunar and planetary instrument. Right: The simple utility of a Newtonian reflector on a Dobsonian mount has made these inexpensive telescopes a popular choice for observers on a budget.

in aperture and greater were becoming increasingly popular.

Dobsonian mounts are simple wood structures with Teflon bearings that provide smooth vertical/horizontal motion. Their simplicity translates into lighter, more compact mounts with surprising stability. Named after California amateur astronomer John Dobson, who popularized the design, the Dobsonian has become the preferred mount for large Newtonian reflectors and a solid choice for smaller models too.

Many observers consider the Dobsonian's inability to track a celestial target automatically (it must be manually recentered about once a minute) as a fair price to pay in exchange for portability. Amateur astronomers are now wielding telescopes up to 25 inches in aperture on these simplified pedestals. These large telescopes are sometimes referred to as "light buckets," because they collect far more starlight than do their smaller cousins. However, they are strictly deep-sky instruments, suitable for galaxies, nebulas and star clusters; they seldom outperform smaller telescopes for viewing the Moon, planets and multiple stars. But one look at a globular cluster or a galaxy through a light bucket, and it is hard to deny the lure of these large telescopes.

In addition to the introduction of Dobsonian mounts, the 1970s saw an even more important revolution in amateur telescopes: the mass production of Schmidt-Cassegrain systems. A **Schmidt-Cassegrain** telescope combines many of the best features of the refractor and the Newtonian reflector. It has a concave main mirror like the Newtonian and a lens at the top of the tube that performs the triple function of correcting for optical aberrations, sealing the tube from dust and other airborne pollutants and supporting a second mirror which reflects the concentrated light back through a hole in the main mirror. The light is finally focused at the rear of the telescope.

The primary advantage of a Schmidt-Cassegrain is the instrument's extremely compact configuration. The tube of an 8-inch is about two feet long, compared with five feet for a

Above: Immediately after its commercial introduction in the early 1970s, the medium-aperture Schmidt-Cassegrain telescope became a strong favorite among amateur astronomers. Its image as an ideal all-round astronomical telescope was somewhat tarnished in the 1980s, however, when large numbers were produced with mediocre (or worse) optics. But today's Schmidt-Cassegrains are fine performers, their compact tubes and fork mounts making them the most portable of all telescope designs.

An Ideal Beginner's Telescope

There may not be a perfect telescope for the beginner, but the closest thing to it is the 6-inch Dobsonian-mounted Newtonian reflector. These telescopes offer the best combination of modest price, versatility and practicality available in a commercial telescope. For about $500, you get a complete telescope of astonishing capability. It will reveal Cassini's division in Saturn's rings, Jupiter's red spot, the polar caps and dark regions on Mars, thousands of lunar features, the Trapezium at the core of the Orion Nebula, hundreds of stars in the Hercules cluster and galaxies 60 million light-years away.

Other telescopes will do this too, but not in this price range. Moreover, the 6-inch Dob is the product of a unique combination of factors that, in my opinion, make it an overachiever in the telescope world and the ideal beginner's instrument. First, it comes in two easy-to-handle pieces: four-foot tube and stool-sized mount. Together, they weigh only 30 pounds, making transport and setup a snap. Second, the 6-inch f/8 primary mirror is easier to manufacture to the necessary precision than are the mirrors of larger instruments of this type, so the average purchaser is virtually assured of getting a telescope with good to excellent optics. Third, the modest 6-inch aperture is big enough to show all the major categories of celestial objects. In many ways, it is the optimum aperture for a Newtonian, because the primary mirror quickly cools down to ambient temperature and is less subject to tube currents than larger models are. Fourth, the optics in an f/8 Newtonian are easier to collimate than in shorter-focal-ratio telescopes, which include all larger commercial Newtonians.

In short, the 6-inch Dob has a lot going for it. Although these telescopes are reasonably popular, I'm surprised that I don't see more of them in the hands of novice astronomers. Maybe it's because they don't conform to the average person's notion of what a real telescope should look like.

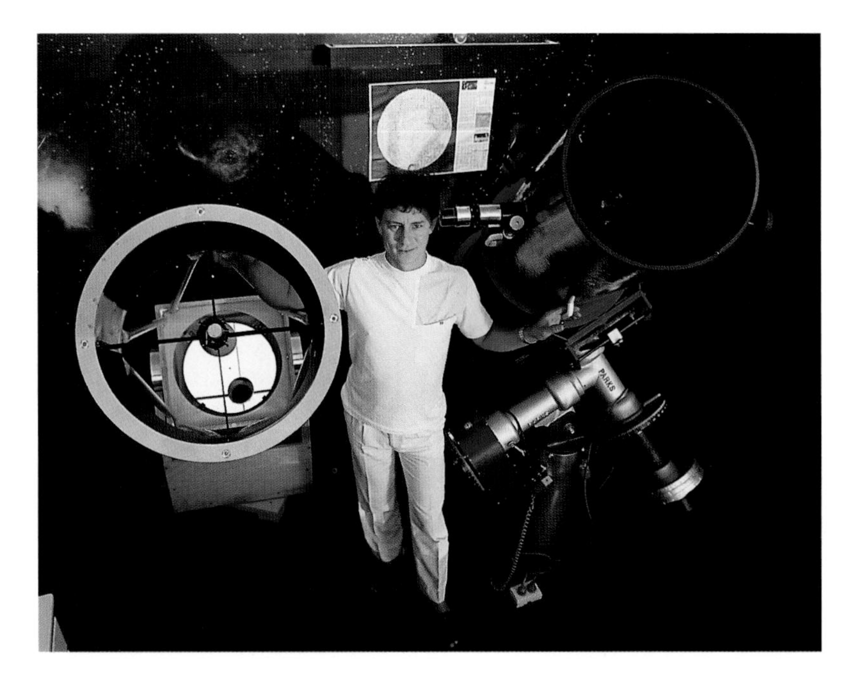

typical 8-inch Newtonian. More telescope is compressed into a smaller package, and that is the main reason the Schmidt-Cassegrain has emerged as the most popular type for serious backyard astronomers. Celestron and Meade are the big-two American manufacturers of Schmidt-Cassegrain telescopes, and both offer excellent value for the money. Eight-inch Schmidt-Cassegrains are priced from $1,000 to $4,000.

Schmidt-Cassegrains cost somewhat more than Newtonians of comparable aperture, especially if the Newtonian is outfitted with a Dobsonian mount, so there are some trade-offs between the two types. But for portability—and this is a major consideration for many backyard astronomers—the Schmidt-Cassegrain is second to none.

A variation of the Schmidt-Cassegrain form, which actually predates it as amateur equipment, is the Maksutov-Cassegrain. For many years, the best-known Maksutov-Cassegrain was the Questar, a beautiful (and very expensive) high-performance 90mm telescope. In the mid-1990s, Meade introduced the ETX, a 90mm Mak-Cass telescope superficially similar to the Questar but much more reasonably priced, at about $600. The ETX is no Questar, but at one-sixth the price, it is an excellent value, especially for those who place a premium on portability.

Making the Choice

What type and size of telescope is the best all-purpose buy? It would be nice if there were an easy answer to this question, but there isn't. Some expert observers will advise you to go for the biggest Newtonian your wallet will allow. Others will counsel you to purchase only the highest-quality optics. Still others will say that portability counts the most. There is no such thing as the perfect telescope for everyone. People are different. They have different lifestyles and interests, not to mention significant variations in disposable income.

I have owned and used dozens of instruments, ranging from small refractors to giant Newtonian reflectors. Personally, I prefer telescopes in the moderate-aperture range (4 to 8 inches). Their ease of use, portability and ability to show all the main types of celestial objects make them the workhorses of amateur astronomy.

I have owned, and since disposed of, several larger instruments, including those I call the hernia models, which are convenient to set up only when two people are avail-

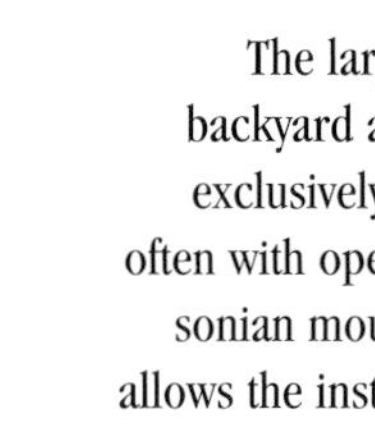

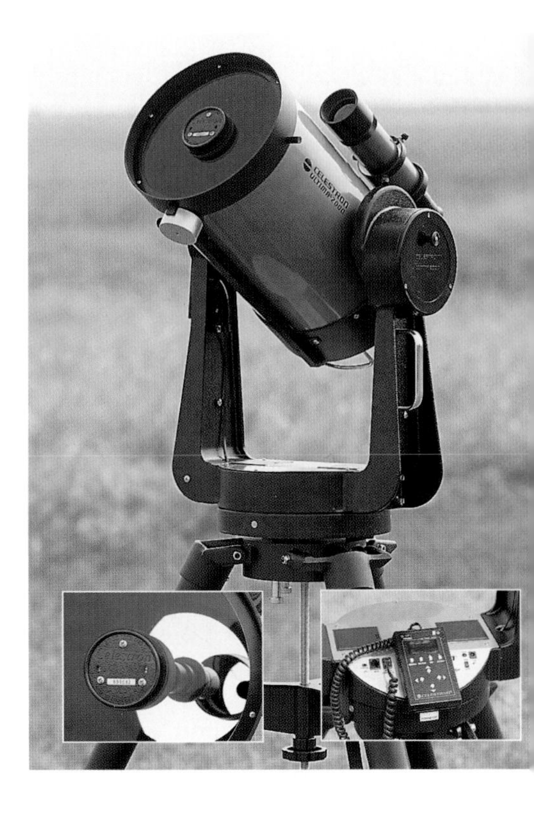

The largest telescopes used by backyard astronomers are almost exclusively the Newtonian design, often with open-strut tubes and Dobsonian mounts. This configuration allows the instruments to house huge primary mirrors, such as the colossal 30-inch at left and the more modest 18-inch at top (left). But for most people, these monsters are simply too big to be practical, regardless of their impressive views of galaxies and nebulas. For the observer who needs a telescope that will fit into a car trunk, the 8-inch Schmidt-Cassegrain is a popular choice. The model above, Celestron's Ultima 2000, has a built-in computer to locate sky objects automatically. Inset photographs show the optical configuration and a close-up of the mount and computer controls.

able, and the Marquis de Sade models, which come in several bulky pieces and have a plethora of bolts and clamps that must be manipulated while you support a huge tube or counterweight. Lack of portability has relegated many such unwieldy models to enthusiasts' storage closets. Aperture fever is a common affliction among amateur astronomers. But until you actually handle telescopes, it is difficult to visualize how much bulkier instruments are in 10-inch and larger sizes. Once the initial euphoria wears off, it can become extremely tiresome to lug out one of those cumbersome monsters. Even some of the stripped-down Newtonian telescopes on Dobsonian mounts are still hefty pieces of equipment.

Size Versus Seeing

Apart from lack of portability, there is another reason why monster telescopes may not perform to expectations. Large optics of excellent quality must be made by hand, an expensive and time-consuming procedure. A big, inexpensive telescope cannot possibly have high-precision optics; consequently, the images may be brighter but are often fuzzier. Second, the larger the telescope, the more likely it is to be affected by a phenomenon known as poor seeing. *Seeing* refers to the steadiness of the image of a celestial object viewed through the telescope. (Good seeing = steady image; poor seeing = unsteady image.)

Turbulence in the Earth's atmosphere imparts a shimmering quality to telescopic images. The intensity of the turbulence varies depending on winds, temperature differential among upper-atmospheric layers, local topography and air circulation immediately around the telescope. The larger the telescope, the worse it will be affected, because large telescopes have to peer through more air than smaller ones do. For example, a telescope with a main mirror or lens 8 inches in diameter must look through a column of air 8 inches wide and about 10 miles long, the nominal thickness of the turbulent layers of the Earth's atmosphere.

Large telescopes are also plagued by "cool-down time," the interval, usually shortly after sundown, during which the optics are shedding heat and their optical figure is slightly distorted. Telescopic images then have a boiling appearance. The cool-down factor is even more prominent when a telescope is taken from the house into the colder air outdoors. Some instruments require hours to stabilize.

However, a large telescope will always produce a brighter image. Light-collecting ability varies with the square of the aperture. A 12-inch telescope produces images nine times brighter than a 4-inch. Therefore, objects nine times fainter can be seen. A 4-inch telescope will reach 13th magnitude, an 8-inch 14th and a 12-inch will probe to 15th magnitude. This advantage is best applied to faint objects like nebulas and galaxies, which show up far better in larger telescopes. But even in these cases, the effects of poor seeing take a toll. Furthermore, a black sky is essential for such deep-sky observing, no matter what instrument is used. If the telescope has to be transported to a dark site, its size can become a liability.

All these factors explain why the most popular telescopes are in the 4-to-8-inch range, the ideal size for portability and

When astronomy enthusiasts set up several telescopes for a night of celestial viewing, there will inevitably be plenty of shared experiences as observers exchange views through the various instruments. Each type and size of telescope has its strengths.

the best compromise between large telescopes, which are somewhat handicapped by sheer bulk and by the effects of poor seeing, and small instruments, whose images are sharp but dim.

Recommendations

What is the bottom line? Begin with a realistic estimate of the maximum you can afford to spend. That, more than anything, will determine your options.

If your budget is less than $500, my top choice would be a 6-inch Dobsonian-mounted Newtonian from Celestron, Orion or Meade. Add a larger finderscope and an extra eyepiece or two, if you can afford them. The 6-inch Dobsonian is the classic beginner's telescope, capable of revealing terrific detail on the Moon and planets and affording pleasing views of nebulas and galaxies. For more portability, the Edmund Astroscan—a 4-inch short-focus Newtonian—or an 80mm short-focus refractor on a camera tripod are possibilities. Most 80mm long-focus refractors or 4.5-inch Newtonians in this class are fitted with inadequate equatorial mounts, but if you find one with a solid mount, such as Celestron's C114HD, it could be a choice as well.

In the $500 to $1,000 category, I suggest an 8- or a 10-inch Dobsonian-mounted Newtonian reflector or an 80mm or a 90mm achromatic refractor on a *sturdy* altazimuth or equatorial mount. Regardless of the type of mount, look for simplicity and functionality. Does the telescope move smoothly and stay *exactly* where you leave it when you take your hand away? It should. If portability is paramount, the Meade ETX 90mm Maksutov-Cassegrain is a good choice. In each case, you will want to start with two or three eyepieces and a Barlow lens (see page 77 for details), which will add to the base price of the instrument, but the package should still be under $1,000.

In the $1,000 to $1,500 price range, Celestron's and Meade's introductory-level 8-inch Schmidt-Cassegrain telescopes are strong contenders. Although these models do not have the computer database and auto-targeting functions of their more expensive siblings, the optics are exactly the same. For greater portability, consider Celestron's C5+, a well-designed 5-inch Schmidt-Cassegrain. Refractor buffs can choose from Meade's and Celestron's 4-inch achromatic models on equatorial mounts.

If you can move into the $1,500 to $2,500 category, several more advanced 8-inch Schmidt-Cassegrain models from Celestron and Meade should get serious consideration. Large Dobsonians are a possibility, but take a good look at 12-inch and larger Dobsonians in person before committing to a purchase. Some of them are gigantic, requiring two people for setup, and once they are in position, a stepladder may be needed to reach the eyepiece. Aperture fever can be a dangerous affliction.

At the opposite end of the scale, 4-inch apochromatic refractors offered by Meade, Tele Vue and Astro-Physics are a popular choice for city and suburban astronomers, who are often restricted by surrounding lights to viewing brighter celestial objects. Apochromatic refractors are prized for their sharp images of the Moon and planets, targets that are often just as plainly seen from urban areas as they are from dark country sites. Of course, many other instruments are available. For more details and telescope comparisons, refer to *The Backyard Astronomer's Guide*, which I coauthored with Alan Dyer.

So far, I have not mentioned building a telescope. Up until about a generation ago, commercial telescopes were a luxury item, substantially more expensive in terms of personal income than is the case now. Active participation in amateur astronomy usually meant becoming a telescope maker. Today, traditional telescope making—mirror grinding and the construction of equatorial mounts—is virtually extinct. Now, almost everyone serious about astronomy buys a complete telescope from a specialty dealer or through a mail-order house.

The only category of telescope making that remains mod-

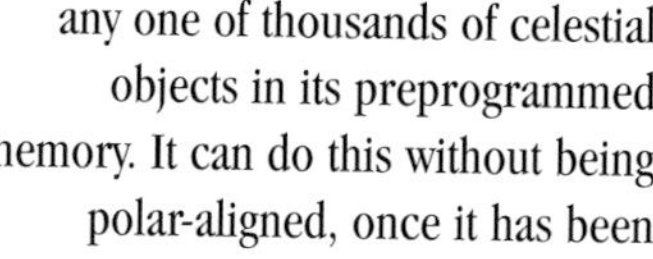

The fully computerized Meade LX200 8-inch Schmidt-Cassegrain, left, is capable of automatically pointing at any one of thousands of celestial objects in its preprogrammed memory. It can do this without being polar-aligned, once it has been initialized by manual sightings on two known stars. At about $3,000, it is a popular choice for the serious backyard astronomer. The ultimate in compact telescopes, the Meade ETX 90mm Maksutov-Cassegrain, above, is just over a foot long and is fitted with a petite fork equatorial mount with motor drive. To polar-align a fork mount, adjust the tripod so that the fork tines are aimed at the celestial pole. Facing page: Composite photograph of observers silhouetted against the Scorpius Milky Way.

erately active is the construction of Dobsonian mounts and tube assemblies, which are then fitted with commercial optics and accessories. Dobsonian mounts are almost entirely wood, which can be worked successfully by any do-it-yourself buff. There was a day when making your own telescope was a way to save a pile of money, but not anymore. Complete telescopes are now at their lowest relative prices ever. Besides, telescope making requires weeks or months of part-time labor. I would sooner spend my time under the stars.

Telescope Accessories

Most commercial telescopes have long lists of optional accessories. In some cases, even the tripod is listed as "optional." Amateur astronomers usually buy a minimum of options initially, adding items as required. This is a wise policy. It allows the owner to get used to the equipment, recognize its deficiencies and evaluate the need for various accessories. Here is a rundown of the telescope accessories most often requested by new telescope owners.

Solar filters: Priced from $50 to $200, these highly desirable aids attach to the front of the telescope to reduce solar radiation to suitable levels for direct viewing of the Sun (see Chapter 8).

Eyepiece filters: Colored glass filters that screw into the base of an eyepiece are said to improve contrast and enhance detail on the Moon and planets. True, but the enhancement is subtle, and these filters can be added to your accessory box as your observing skills develop and you identify a real need for them. Having said that, there is one—the lunar filter—that almost every telescope needs. Lunar filters are neutral gray; they simply reduce the glare and make Moon viewing more comfortable. About $15.

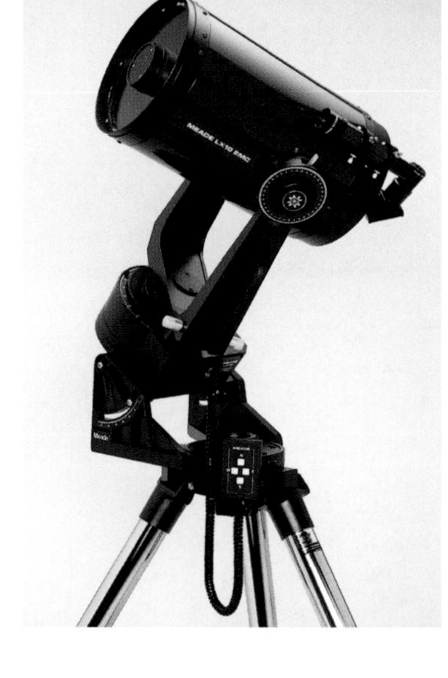

Nebula filters: The glow from cities, shopping centers and streetlights dims the natural beauty of the night sky. A nebula filter attached to the telescope's eyepiece blocks much of the interfering glow, producing better contrast in views of nebulas. Priced from about $60 each, they work best on short-focal-ratio telescopes. Nebula filters provide a minimal advantage on objects other than nebulas.

Slow-motion-control motors: Many equatorial mounts already offer these as standard equipment. You simply push buttons on a control paddle to center an object in the eyepiece field of view. Generally a useful aid.

Telecompressors: These are Schmidt-Cassegrain accessories for low-power deep-sky sweeping and photography. The lens intercepts the light before it enters the eyepiece, widening the field of view. From $80 to $250.

Erecting prisms: It comes as a surprise to most people looking through an astronomical telescope for the first time that the image is inverted (or reversed left to right if a diagonal is in the focuser). But that's the way optics work. To flip the image right side up, an erecting prism, like the ones sealed inside your binoculars, must be positioned in the focuser ahead of the eyepiece. Erecting prisms may sound like mandatory equipment, but in reality, they are never used for astronomy. Adding that extra glass reduces the instrument's light transmission and introduces optical aberrations that become evident at higher power. Astronomers always opt for the optically superior view, not the right-side-up one. The only practical application for erecting prisms is on small refractors used for land viewing.

Dew caps: Dew forming on the front lens plagues owners of Schmidt-Cassegrain and Maksutov-Cassegrain telescopes. Refractors are affected as well. An extension of the telescope tube, known as a dew cap, tends to prevent dew formation. Another solution is to use a hand-held hair dryer to evaporate the dew.

Dew-prevention heaters: If you live in a dew-prone en-

The base model of an 8-inch Schmidt-Cassegrain telescope, top right, comes with an equatorial fork mount, although the tripod is an "option." When the tripod is added, along with other desirable options such as a larger finder and declination slow-motion motor, the price is about $1,500—a good value for this class of instrument. Full-aperture solar filters are the most comfortable and the safest way to observe the Sun. They are available in metallic-coated glass, center, or aluminum-coated Mylar (page 144). Colored eyepiece filters for lunar and planetary observing, bottom, are relatively inexpensive but are not a necessary accessory for the casual observer.

"How Powerful Is It?"

Whenever I demonstrate one of my telescopes, I am invariably asked, "How much did it cost?" and "How powerful is it?" For anyone contemplating a telescope purchase, the cost is certainly an important consideration. But the question about power focuses on one of the most misunderstood aspects of telescope performance. Claims of huge power capabilities are almost totally meaningless.

There are three distinct types of telescopic power—light gathering, magnifying and resolving. Least important is magnification, yet magnifying power alone is often used as a selling point for small telescopes. The most significant ingredient is light-gathering ability. For example, the Orion Nebula will look about the same size in a 50mm telescope as it does in a 100mm if each is used at the same magnification. However, the light-gathering power of the 100mm telescope is four times greater than the 50mm, so the nebula will be four times brighter.

This is an important difference, since most astronomical objects are relatively faint and need their brightness significantly boosted. The images must be bright before they can be subjected to magnification. Only the Sun, Moon and brighter planets are sufficiently luminous that light-gathering power is not a crucial ingredient for a decent view.

Given adequate light-gathering ability, there are other limits on magnification. As a general rule, the practical maximum magnification limit is about 2x per millimeter (50x per inch) of aperture. This means that the upper limit for a standard 60mm refractor is 120x. When such a telescope is pushed to 200x, the images are grossly overamplified and exceed the third factor on our list: resolving power—the ability of the instrument to discriminate fine detail. The limitation on resolving power is imposed by the interaction of light and optics.

Excessive magnification not only exceeds the telescope's capabilities by producing grossly fuzzed images but also makes the instrument almost impossible to use. Tiny jiggles created when the focusing knob is touched or movements generated by a breath of wind become amplified by the same amount that the instrument is magnifying, causing the star or planet to quiver or lurch across the field of view. A further drawback is the tiny eye-lens opening in the eyepieces required to reach high powers. Sometimes, it is like trying to peer through a pinhole. The narrow field of view means that more time must be spent locating and centering sky objects. And, once located, celestial objects soon drift out of the field of view due to the Earth's rotation (unless the telescope is equipped with an equatorial mount and drive).

And if all this has not discouraged the power-hungry tyro, then the ubiquitous problem of atmospheric turbulence and its frequent poor-seeing conditions will be the final damper. The 2x-per-millimeter limit applies only in good seeing, when air turbulence is minimal. Telescopes are often seeing-limited to half that.

Normal operating magnification for astronomy is one-tenth to one-half the 2x-per-millimeter maximum. This range offers the best ratio between aperture and magnification by providing the correct balance of light-collecting and magnifying power. On my 6-inch telescope, for example, I rarely use magnifications over 200x, and 40x provides the most stunning views of brighter star clusters.

If telescope advertising trumpets power capabilities beyond the limits of even the best possible optical systems, look elsewhere.

Magnification, or power, is only one component of telescope performance, and high power alone is no indication of a telescope's capabilities. Even large telescopes are seldom used at magnifications above 250x.

vironment, these low-voltage heaters that wrap around strategic parts of your telescope will keep your optics dry all night. The low level of heating does not affect the telescope's performance. Price varies with number and size of heaters. A typical setup will cost from $100 to $200.

Photographic accessories are definitely best left until you have had some experience with the telescope in straightforward visual observing. Avoid the temptation to load up on adapters, off-axis guiders and dual-axis slow-motion motors until you are fully familiar with the comparative difficulties involved in various types of astrophotography. Most of the photographic techniques suggested in Chapter 11 require no adapters and should be tried first before plunging into more difficult aspects of celestial portraiture.

Telrad is a trademark for a unique one-power bull's-eye sighting device that makes a dandy finderscope.

Motorized focusers are useful on telescopes that quiver when you touch the focus knob—most telescopes. However, this accessory is highly personal. Some observers can't be without push-button focusing; others prefer the manual method.

Computer controls that automatically point your telescope toward any one of thousands of celestial objects in the computer's memory banks when you simply key in the object's name sound like just the thing when you are finding your way around the sky. The trouble is, you will never learn anything if the computer is doing all the finding. For more on this, see "Computer-Age Scopes" (page 79).

Eyepieces

Sometimes, telescopes with otherwise excellent optics are supplied with bottom-of-the-line eyepieces. A good optical system will perform to capacity only when linked to top-quality eyepieces. Inexpensive beginners' telescopes are often equipped with the old Japanese standard-sized eyepieces—0.965-inch outside diameter (the width of the section that slips into the eyepiece holder on the telescope). The American standard is 1.25 inches. If a telescope has a 0.965-inch-diameter eyepiece, the only solution for better performance is to change the focusing mechanism to the higher diameter or to buy either a 1.25-inch adapter or a hybrid diagonal (0.965 inch on one end, 1.25 inches on the eyepiece end). Even some fairly high-priced telescopes are not supplied with top-of-the-line eyepieces.

Just as there are different types of telescopes with differing optical configurations, eyepieces (sometimes called oculars) are available in a range of types, each with its own advantages and disadvantages. Here is what you need to know to make informed choices:

Huygenian and *Ramsden*, the simplest eyepiece types available, have been around for centuries and are usually found in 0.965-inch sizes. They accompany only the cheapest telescopes, since they are worth just a few dollars each. They can be identified by the letters R, H or AR engraved on the outside. Anyone who owns an inexpensive telescope undoubtedly has one or two of these. They have been completely superseded by the other designs.

I also rank *zoom* eyepieces in the undesirable category for astronomical use. They seem like a good idea, but in practice, the field of view is restricted and the optical performance is generally inferior.

The *Kellner* eyepiece is also an older design, dating back to the 19th century, but it is still in use today on refractors and Schmidt-Cassegrains and provides acceptable performance for low- or medium-power applications. Modest price is its prime attribute. Its main disadvantage is a relatively narrow field of view. The RKE series of eyepieces, developed by Edmund Scientific in 1978, is an improvement on the Kellner design. The RKE reverses the arrangement of the three lenses in the eyepiece and utilizes modern high-index glass. Other manufacturers followed suit with similar designs under different names, all of which yield better imagery and a wider field than the standard Kellner and at a relatively low price, usually less than $60.

Orthoscopic eyepieces were once regarded as the finest available, and for many years, they were. They improve on the Kellner design by adding a fourth lens element, which provides a wider field of view and eliminates virtually all optical aberrations. Price range is $60 to $120. Orthos are excellent eyepieces, still preferred by some backyard astronomers for medium- and high-power applications, especially planetary observing.

Plössl eyepieces are in many ways superior to the types men-

The small black box on the telescope at left is the Telrad, a one-power sighting device that ingeniously "projects" a red bull's-eye on the sky for easy aiming. Similar but more compact projection finders are the Rigel Systems' Quick-Finder and TeleVue's Quik-Point. Many backyard astronomers prefer projection sights over a normal finder or use both in conjunction. Above: A 0.965-inch and a 1.25-inch eyepiece. The latter is the North American standard for telescope eyepieces.

tioned so far. Their four-element design provides remarkably sharp images with a slightly wider field of view than the Orthoscopic eyepieces. Plössls are extremely versatile, being suited to low-, medium- and high-power applications. Prices range from $60 to $120. I prefer Plössls for many applications, and for observers on a budget, "house-brand" Plössls are offered by several telescope companies for less than $50.

Erfle eyepieces became popular after World War II because the five-element design was widely used in military optics. Erfles have long been favored for low-power applications because of their extremely wide field of view. By today's standards, however, these eyepieces are outperformed by the more modern wide-field designs described below.

Modern *wide-field* eyepiece designs with six or seven elements are a recent advance on the Erfle design and offer better imagery across the full field. They are very expensive, especially in the lowest-power versions that require a lot of glass. The newest wide-field design, *Panoptic* by Tele Vue, is the premium entry in this class, offering superb low-power definition.

Another elegant eyepiece design is the *Nagler* series introduced in 1981 by Tele Vue, undoubtedly the finest medium- and high-power eyepieces available. A similar design, called *Ultra Wide Angle*, was introduced by Meade a few years later. The eyepieces cost as much as a small telescope ($200 and up), but their razor-sharp, extremely wide fields of view noticeably improve the performance of any telescope. However, all that glass means that these are heavy eyepieces which can overweigh a small or delicately balanced telescope.

The magnification provided by an eyepiece can be determined by dividing the focal length of the telescope by the focal length of the eyepiece. A 2000mm-focal-length telescope using a 25mm eyepiece will yield 2,000 ÷ 25 = 80x. The eyepiece focal length is always engraved on the side, along with the type of eyepiece or a letter indicating the type (e.g., K25mm = Kellner 25mm focal length). Low-power eyepieces range from 40mm to 20mm focal length; medium power from 19mm to 13mm; high power from 12mm to 4mm. I recommend having at least one of each or a low- and medium-power eyepiece and a Barlow lens.

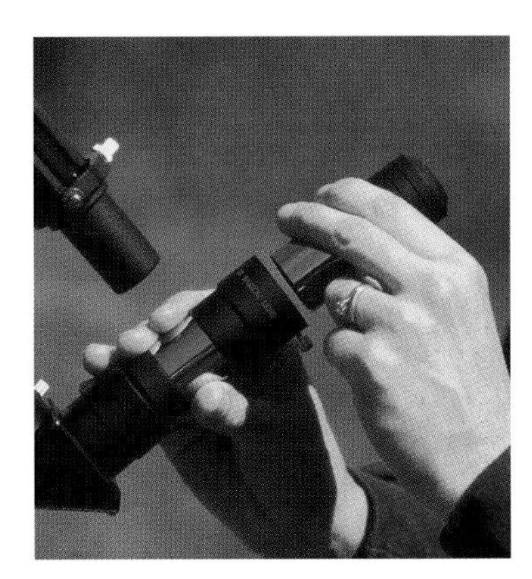

A *Barlow* is basically a tube with a lens at the lower end and fittings to accept an eyepiece at the upper end. A Barlow amplifies the magnification of the eyepiece two to three times, depending on the manufacturer's specifications. I almost never use eyepieces shorter than 12mm focal length because they often have uncomfortably short eye relief—a term that refers to the ideal distance the eye must be from the eyepiece to take in the full field of view. A 2x Barlow and a 16mm eyepiece are more comfortable to use than an 8mm eyepiece alone. Both yield exactly the same magnification, but a 16mm eyepiece has significantly greater eye relief than an 8mm.

A typical Barlow costs about the same as a good-quality eyepiece. Beware of inferior-quality Barlows supplied as standard equipment on some telescopes. In 1998, Tele Vue introduced *Powermate*, a 5x Barlow with a more sophisticated design than previous Barlows (at twice the price). The Powermate works very well but is a special-purpose accessory suited for use with telescopes that have excellent optics and a focal length of less than 1000mm.

Above (left to right): The basic eyepiece sizes are 0.965-inch, 1.25-inch, 2-inch and dual 1.25/2-inch. The dimension refers to the diameter of the barrel that slips into the telescope focuser. The 0.965-inch size is found only on some inexpensive beginners' telescopes. The 2-inch size gathers a larger cone of light for optimum wide-field low-power viewing.
Top right: Diagonals in both 1.25- and 2-inch sizes are available for comfortable viewing with refractors and Schmidt-Cassegrains. Top left: Barlows are designed to go into the focuser directly ahead of the eyepiece. If placed ahead of the diagonal, a Barlow will amplify about $1\frac{1}{2}$ times its rated value, but the telescope's focus point may be pushed to a position beyond the focuser's travel limit.
Above center: A motorized focuser can eliminate the shakes caused by hand-focusing. The utility of this accessory depends on the stability of the individual telescope and on the owner's personal preference.

Many large telescopes can accommodate *2-inch-diameter eyepieces*, which, although they may cost more than $200, can offer stunning low-power performance with fields of view substantially wider than any 1.25-inch eyepiece.

Eyepieces are important. Quality eyepieces will improve any telescope's performance. Your eyepiece collection should be worth at least one-third of the cost of your telescope. Eyepieces have excellent resale value and can be used on any telescope, so your collection has permanent value.

With hundreds of eyepieces now available, the decision about what to buy is tougher than ever. I have used many of them, and here are some of my favorites: 35mm Panoptic by Tele Vue (2-inch barrel); 32mm to 15mm Plössls by Meade and Tele Vue; 22mm Panoptic by Tele Vue; 21mm RKE by Edmund; 18mm Super Wide by Meade; 16mm König by University; 16mm Nagler by Tele Vue; 13.8mm Super Wide by Meade; 9mm and 7mm Naglers by Tele Vue; and 8.8mm and 6.7mm Ultra Wides (Nagler type) by Meade. I also like the 2x Ultima Barlow by Celestron and the 3x Tele Vue Barlow. You will never need all of these, and there are many other fine eyepieces on the market besides those mentioned.

Although I wear eyeglasses, I seldom use them when at the eyepiece. A slight change in focus accommodates all vision differences except astigmatism. (Those who wear glasses to correct astigmatism should leave them on.) Most eyepieces, especially medium- and high-power, require the eye to be closer to the lens than glasses permit. Keeping glasses on reduces the field of view but otherwise does not substantially alter the image.

Focal Ratio

Focal ratio is something you will need to know about both before and after you become a telescope owner. Every telescope operates at a specific focal ratio. The focal ratio is usually marked on the instrument or supplied in the owner's manual. Examples are f/10 and f/4.5. The focal ratio is determined by dividing the telescope's focal length (the distance that the main lens or mirror refracts or reflects light to the point of focus) by the diameter of the main lens or mirror. Thus an 80mm refractor with a 1200mm focal length is f/15, and a 10-inch Newtonian with a 45-inch focal length is f/4.5.

Newtonian reflectors and refractors are about as long as their focal lengths, so just measuring the length of the tube and dividing by the aperture gives a rough focal ratio. Not so with Schmidt-Cassegrains and Maksutov-Cassegrains. These systems fold the light path and squeeze a long focal length into a short tube. Schmidt-Cassegrains are usually f/10, although some are f/6.3. Maksutov-Cassegrains range from f/7 to f/16. Newtonians vary from f/4.5 to f/10, and refractors are f/11 to f/16 (standard, or achromatic, versions) or f/6 to f/9 (apochromatic). For a variety of reasons, the smaller the focal ratio, the more difficult it is to manufacture good optics.

One persistent myth is that telescopes with small focal ratios—sometimes called "short" or "fast" focal ratios—produce brighter images than do telescopes with large focal ratios. This is true *only* for photography through the instrument. Visually, any two telescopes of a given type and aperture (say, two 8-inch Schmidt-Cassegrains, one f/6.3, the other f/10) will produce identically bright images *when used at the same magnification*, even though they have different focal ratios. What is different is that telescopes with short focal ratios are capable of operating at lower minimum power. However, in most instances, f/6 is short enough for all visual applications. There is no ideal focal ratio, although I have often found that the most versatile telescopes of all types are f/6 to f/8. (Exception: large Newtonian reflectors where f/5 optics keep the tube a manageable length.)

Above: Modern computer software can be used to control a telescope by keyboard and mouse. The observer simply clicks on the object displayed on the monitor, and the telescope moves to the same object in the sky. The owner then commands the CCD camera on the telescope to take an image of the object, which is, in turn, displayed on the monitor. Some would consider this the height of astronomical decadence, since the "observer" isn't even looking at the real sky. But today's technology has given amateur astronomy a wider appeal than ever before *because* the possibilities are so vast. The choice is yours. You can enjoy it as a technofest or purely as a relaxing diversion extending no further than binocular stargazing—or anything in between.

Computer-Age Scopes

Many telescopes with equatorial mounts are equipped with two numbered dials, called setting circles, on the rotational axes. These are for correlating the aiming of the telescope with the sky coordinate system, which is analogous to latitude and longitude on Earth. A celestial equator and pole, directly above their Earthly counterparts, are the keys to the grid system.

Navigating this celestial-coordinate system with setting circles may seem like the logical way to track down sky objects, yet relatively few amateur astronomers use this method. Significantly, the more experienced the backyard astronomer, the less the setting circles are used. To use the setting circles, the observer (1) polar-aligns the telescope, (2) selects a bright star that is as close as possible to the area containing the target object and centers it in the telescope's field of view, (3) looks up the coordinates, known as right ascension and declination, of the centered star, (4) adjusts the dials to reflect the star's coordinates, (5) looks up the coordinates of the target object, (6) turns the telescope so that the setting circles are at the target's coordinates and, finally, (7) peers through the eyepiece to see whether the celestial target is in view.

There are two problems with this, one practical and one philosophical. The practical problem is that the telescope has to be precisely aligned to the celestial pole and the setting circles must be accurate. With practice, aligning the telescope does not take much time, but the setting circles on many amateur telescopes are not precise enough to get the desired object in the field of view every time. Recently, however, the computer age has surmounted this shortcoming with digital-readout setting circles (a few hundred dollars) that are more accurate than mechanical setting circles, are easier to use and can be fitted to most commercial telescopes. Certain models are even programmed to compensate for errors in polar alignment, although the observer must initialize the setup by pointing the telescope at two stars from a list provided by the manufacturer.

Even more sophisticated (and more costly) are computerized telescopes introduced in the 1990s, with the computer built right into the mount. The observer initializes the telescope by pointing it at two bright stars—no polar alignment required. For the rest of the night, the computer tracks the sky to compensate for the Earth's rotation. In addition, the computer has a database of thousands of celestial objects that can be called up by pressing buttons on a control panel. Once an object is selected, the computer uses two high-speed stepper motors in each axis to slue the telescope to the object's position.

Such an arrangement may sound ideal for the beginner who has yet to learn the sky. But this is where I have the philosophical problem. The challenge of hunting down celestial quarry using the eye, finderscope and telescope is pure backyard astronomy. Circumventing the hunt is like wearing a bag over your head while under the stars and removing it only to look in the eyepiece. If experience at tracking down faint galaxies and nebulas is bypassed, the night sky never becomes the comforting dome of familiar pathways that the true backyard astronomer knows it to be. The smudge of a remote galaxy means more when *you* find it, rather than some computer.

I searched for nearly an hour one summer night for the globular cluster M3 when I began tracking down telescopic quarry. Now it takes only a few seconds. However, if I had always been doing it using setting circles or a computer-aided telescope, it would still take the same amount of time working through a basically mechanical ritual. By using the visual-sighting method, I have come to know the sky intimately.

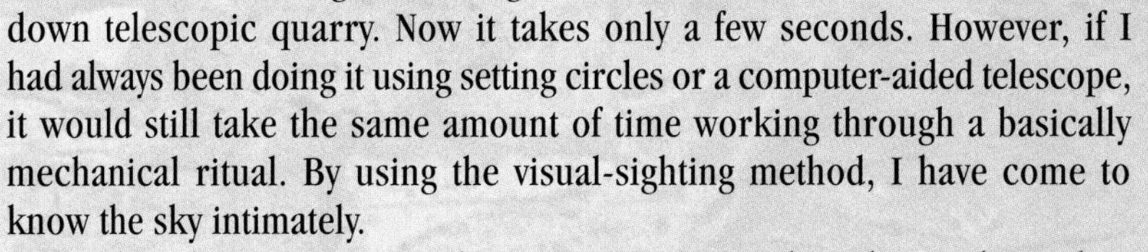

Although I urge novice telescope users not to shortchange themselves by taking the seemingly easy computer-guided route to the galaxies, digital setting circles and computers do have their place. Several of my colleagues, all experienced amateur astronomers, say that these accessories have allowed them to pursue more specialized and productive observing programs. I also acknowledge that for many busy people, time is at a premium. They want to observe the sights on the few nights they do spend at a dark site. So my philosophical point is just that, and it may not be practical for you. But a computerized telescope often costs more than twice as much as a noncomputer version of the same instrument, so weigh your priorities carefully.

Telescope setting circles are graduated in degrees of declination, far left, and hours and minutes of right ascension, left. As tools to zero in on celestial objects, they have been largely superseded by computer-controlled telescopes, above.

TELESCOPE COMPARISONS

Type	Variations	Aperture Range in Common Use	General Specifications/ Performance	Applications
Refractor	achromatic	2.4" to 4"	These rugged, reliable, generally maintenance-free instruments are a traditional "first" telescope for backyard astronomers. Easily portable in 3-inch and smaller sizes. Beware the inadequate mounts and spindly tripods on less expensive models.	Excellent for Moon, planets, star clusters, double stars and general introductory scanning. Poor on nebulas, faint clusters and galaxies due to small apertures. Best buy for city or suburban sky viewing, where faint objects are obscured anyway.
	apochromatic	3" to 7"	Unquestionably the finest telescope optical systems available because of totally aberration- and obstruction-free lens design. Superb imagery. Expensive. Easily portable in 4-inch size.	Outstanding lunar and planetary performance in all sizes. A favorite for serious astrophotographers. Ultrasharp star images and good deep-sky penetration. Limited only by relatively small aperture, compared with Newtonians.
Newtonian Reflector	equatorially mounted	4" to 12"	If well made, can provide excellent value and fine performance, especially in 6-inch size. Requires more maintenance than other types. Cumbersome and not really portable in sizes over 8 inches.	Yields good results on all types of backyard telescopic activities; especially appropriate well away from light-fogged skies.
	Dobsonian-mounted	4" and up	Simplified lightweight mount results in large aperture in a relatively compact and inexpensive package. Very portable in smaller sizes; 6-inch size is especially recommended for beginning astronomers.	Larger models (10-inch and up) intended to be used under dark-sky conditions for nebulas, clusters and galaxies. Smaller models fine for all types of celestial observation. Tracking must be done manually at all times due to simplified mount design.
Compound	Schmidt-Cassegrain	5" to 16"	Optical performance generally good. Stubby tube results in stable, easy-to-use system. More expensive than equatorial Newtonian types, but only modestly so.	Compact design and vast array of accessories for all applications make the Schmidt-Cassegrain a good all-round instrument for backyard astronomy.
	Maksutov-Cassegrain	3.5" to 7"	Well known for excellent performance and compact format but has never dominated the amateur-astronomy scene.	Good for every backyard application except wide-field low-power viewing. Small models are extremely compact; excellent for frequent travelers.

FACTORS TO CONSIDER WHEN SELECTING A FIRST TELESCOPE

Telescope type	Will telescope usually be transported to observing site?	Is main observing site an urban location or a dark rural location?	Will telescope be used for celestial photography?
2.4" to 4" achromatic refractor	Generally easily transportable.	Good performance in urban environment; will not equal other types under dark skies.	Not recommended except for shooting the Moon.
3" to 7" apochromatic refractor	Easily transported up to 5-inch size; larger models are hefty.	A good choice if you are limited to a mix of mostly urban and some rural observing.	A favorite for astrophotographers.
4" to 8" Newtonian reflector, equatorial mount	Easily transported, although some 8-inch units are bulky.	Good all-round performer.	Astrophoto versions often have beefier mounts and specialized accessories.
10" or larger Newtonian reflector, equatorial mount	10-inch units are about the limit for convenient transport.	Performance limited in urban environment; excellent performer under dark skies.	Will yield excellent results, but only with massive mount and proper accessories.
4" to 8" Newtonian reflector, Dobsonian mount	Easily transported.	Good all-round performer.	Not recommended.
10" or larger Newtonian reflector, Dobsonian mount	Transportable, but a van may be needed.	Urban environment severely limits usefulness of this type; designed for use under dark skies.	Not recommended.
4" to 8" Schmidt-Cassegrain	Easily transportable in any vehicle.	Good all-round performer.	Excellent range of accessories for backyard astrophotographers.
3.5" to 7" Maksutov-Cassegrain	Exceptionally compact in smaller apertures.	Often a favorite for urban astronomers who occasionally observe from really dark sites.	Long focal ratio* limits photography to Moon or CCD imaging.

Telescope type	Will telescope be used primarily for viewing Sun, Moon, planets and other bright, easy-to-find objects?	Will telescope be used primarily for viewing nebulas, star clusters and galaxies?	Will telescope frequently be used for daytime land viewing?
2.4" to 4" achromatic refractor	Excellent performance; long focal ratios* and equatorial mount preferred.	Not recommended.	Suitable and recommended.
3" to 7" apochromatic refractor	Unsurpassed for consistently fine lunar and planetary performance. The preferred choice if you can afford one.	Excellent definition and contrast but limited by their aperture.	Tele Vue "Pronto" and Astro-Physics "Traveler" especially recommended for daytime viewing.
4" to 8" Newtonian reflector, equatorial mount	Generally good performance; used to be standard planetary telescope until advent of apochromatic refractors in 1980s.	6- and 8-inch models especially recommended.	Not recommended.
10" or larger Newtonian reflector, equatorial mount	Good results obtained in medium focal ratios* with premium optics.	Top-rated for deep-sky viewing.	Not recommended.
4" to 8" Newtonian reflector, Dobsonian mount	Generally good performance, although mounts not intended for continuous high-power planetary viewing.	6- and 8-inch models especially recommended.	Not recommended.
10" or larger Newtonian reflector, Dobsonian mount	These telescopes are not designed for observing bright objects and seldom produce satisfactory views compared with other types.	Very popular as the most economical way to see the faint fuzzies. Excellent at low powers, which are used most of the time anyway.	Not recommended.
5" to 16" Schmidt-Cassegrain	Although apochromatic refractors are best in this category, the Schmidt-Cassegrain offers good planetary views.	Deep-sky performance rivals that of any other telescope type.	Suitable in smallest sizes.
3.5" to 7" Maksutov-Cassegrain	When well made, this design seems to produce performance second only to refractors.	In most common size (3.5-inch), these instruments are less effective than other types simply because of low light-collecting power.	Ideal in 3.5-inch size.

*Short-focal-ratio telescopes are usually f/4 to f/5; that is, the focal length (effective distance from main optical element to eyepiece) is four or five times the diameter of the objective lens or mirrors. Medium-focal-ratio telescopes are f/6 to f/8; long-focal-ratio, f/9 to f/16.

CHAPTER SIX

PROBING THE DEPTHS

What is inconceivable about the universe is that it should be at all conceivable.

Albert Einstein

An acquaintance who was aware of my interest in astronomy noticed my telescope in the backyard at dusk one evening. "What do you actually *do* with that?" he asked. I launched into an enthusiastic description of the universe of planets, stars and galaxies and the thrill of personal discovery that comes with observing them. After listening attentively, he then wondered what I did after I had seen it all.

I tried to explain that I spend most of my time reexamining celestial objects I have seen before, but I realized how difficult such a concept must seem. Backyard astronomers are a special breed. They savor their moments under the stars. They have an infatuation—a love affair—with the cosmos that grows and nurtures itself just as meaningful human relationships do. Of course, it is a less definable one-way relationship, but I have come to regard that feeling as the closest I can ever come to being at one with nature. After a night under the stars, I have a sense of mellowness, an amalgam of humility, wonder and discovery. The universe is beautiful, in both the visual and the spiritual sense.

The visual beauty is at least clearly definable. It comes in a variety of forms, ranging from stars with heartbeats to villages of suns perched in the Milky Way Galaxy's nearby spiral arms to remote stellar cities, the galaxies. It is now time to be specific, to define exactly what can be seen with a telescope or binoculars and where in the sky to look. In the following chapters, we will examine solar system phenomena—the Sun, planets, their satellites and comets—and nighttime atmospheric phenomena, such as meteors and auroras. But all of that is our cosmic backyard. The universe beyond the solar system, explored in this chapter, is known to amateur astronomers as the deep sky, and the targets of interest are commonly called deep-sky objects.

Even though most stars in the nighttime

This rich sector of the Milky Way in Cygnus gives the impression that stars are almost on top of one another. In reality, they are light-years apart and appear shoulder to shoulder only because we are looking deep into the galaxy's spiral arms. Facing page: The Trifid Nebula in Sagittarius.

sky are bigger and brighter than our Sun, nothing but a tiny point is seen telescopically, no matter how large the instrument or what magnification is used to observe a star. Stars are simply too far away. Under high magnification, a star actually does appear as a disk, but that results from the nature of optics and their interaction with light. It is called the Airy disk, after the 19th-century English astronomer who explained it.

Double Stars

Despite the fact that most stars are solitary dots when viewed through a telescope, single stars are in the minority. At least half and possibly as many as 80 percent of all the stars in the Milky Way Galaxy are members of double- or multiple-star systems—two or more suns gravitationally bound and orbiting about one another. Quite often, these stars orbit more closely than one astronomical unit (AU). To be visible in backyard telescopes, the components of a binary or multiple star must be at least several dozen AU apart. There are a few thousand of these star systems scattered across the sky, some making exquisitely beautiful sights in binoculars and small telescopes.

Sometimes, two stars of a binary are exactly the same brightness, others differ slightly, and still others have wide variations in luminosity that contrast the two suns. However, the prizewinning doubles are the few with stars of completely different colors, indicating a wide difference in surface temperature. I enjoy looking at binary stars, knowing that I am gazing at two suns, each presumably capable of having planets. Earth could have been born into such a system. How different our sky would appear if a small red sun orbited in the place of Neptune.

The apparent distance between the components of double stars is measured in fractions of a degree. Remember that the Moon is about half a degree in diameter. When viewed through a telescope or binoculars, the components of most double and multiple stars appear about as far apart as opposite sides of a small lunar crater. To define this distance, astronomers use minutes (') and seconds (")—one minute is one-sixtieth of a degree, and one second is one-sixtieth of a minute. The unaided eye can distinguish two equally bright stars 6' apart, about one-tenth of a degree. If the stars are closer than that, binoculars are needed. Their increased resolving power will separate doubles down to just under one minute.

A 60mm telescope vastly increases the number of multiple stars visible, because it can separate two equally bright suns apparently only 2.0" apart. A 3-inch telescope under the best conditions will

Using Your Night Eyes

Although humans cannot match the superb night vision of owls, cats and other nocturnal creatures, our eyes are remarkably efficient at seeing detail in the dark. An elaborate system for seeing in the dark evolved millions of years ago, when our ancestors had to perceive potential danger after nightfall. The process is called dark adaptation.

In a darkened environment, the eye reacts almost immediately by increasing the size of the pupil, thus allowing more light into the eye. (This is the same as opening the f-stop on a camera from f8 to f2.) Then, during the next 15 minutes or so, a more complex process takes place. The supply of visual pigment in the photoreceptors of the retina steadily increases, and as long as no bright light enters the eye, the sensitivity to dim light levels surges. This pigment-sensitizing process can be likened to changing the film in a camera from 50-speed to 3200-speed.

Complete dark adaptation has astonishing consequences for stargazers. On a black, moonless night, the Milky Way stands out as a swath of light. To someone stepping out of a normally illuminated house, it is virtually invisible, but to the dark-adapted skywatcher, the Milky Way is transformed into a profusion of faint stars, producing the illusion that the sky is covered with them.

Fully dark-adapted eyes can see about 3,000 stars in a dark sky well away from city lights. That may not seem like many, but compared with the 200 or so seen from typical well-lit suburban areas, it is almost the proverbial difference between night and day.

Experienced stargazers preserve their dark-adapted vision by using low levels of red light when adjusting equipment or consulting charts and reference books. Red light is much less damaging to night vision than the white light of an unfiltered flashlight.

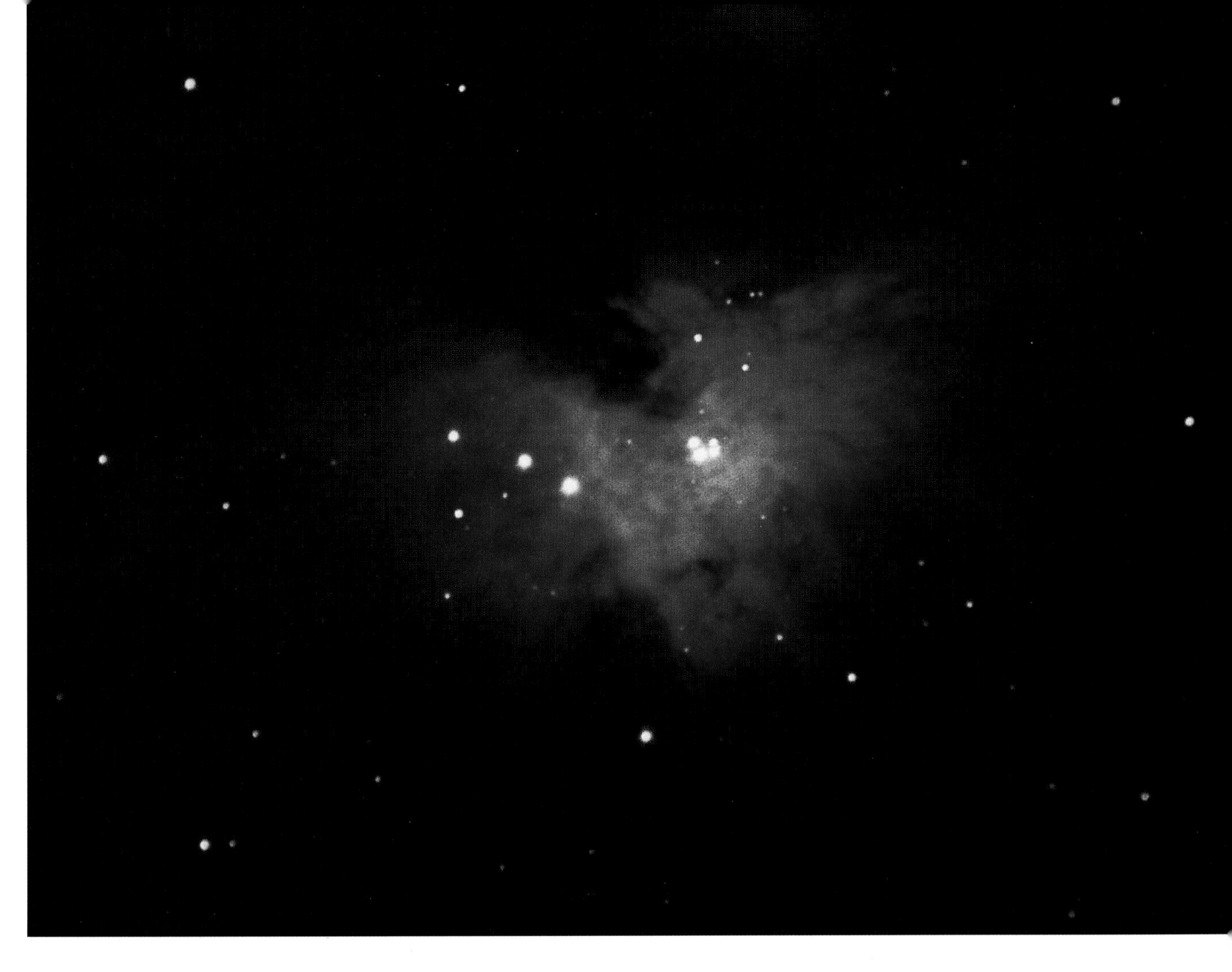

divide two stars 1.5" apart. Using such an instrument, Massachusetts amateur astronomer Glenn Chaple has observed more than 1,400 different double- and multiple-star systems.

About 10,000 double-star systems are visible in a 6-inch telescope, which will resolve two stars of equal brightness 0.8" apart. This is an exceedingly small measure, comparable to the apparent width of a dime five kilometers away. Larger telescopes will resolve proportionally closer pairs. However, these are theoretical figures assuming excellent optics *and* perfect atmospheric conditions. And, most important, the limits also assume that an experienced observer is at the eyepiece.

Beginners should not expect to come close to the telescope's theoretical limit, at least at first. And, indeed, that is not the point of double-star observing. Stars that are easily distinguished make prettier sights than the ones that are so close together the images can barely be separated. With these factors in mind, I have selected the best double and multiple stars for typical amateur equipment—both binoculars and telescopes. These stars are specifically identified on the charts at the end of this chapter. In some cases, the actual distance between components is given along with the apparent distance.

Not all stars that appear as doubles are together in space. A chance alignment of stars is called an optical double. The technical designation for a pair of gravitationally associated stars is a binary system, but they are usually simply called double stars.

The brightness of individual stars in binary systems is usually given to an accuracy of one-tenth of a magnitude. Initially, it may seem difficult to tell a second-magnitude star from a third-magnitude star, but with a little practice, it is possible to distinguish stars differing by only a few tenths of a magnitude, especially when they are fairly close to one another. Many of the naked-eye stars on the charts in this chapter have their magnitudes indicated to the nearest tenth. After a few nights' practice, it should not be difficult to see the difference between, for example, a star of magnitude 3.3 and one of 2.9. In addition to their magnitudes, the separation of the stars in a multiple-star system is tabulated in minutes or seconds. The brightest star is known as A, the second brightest B, the third brightest C, and so on.

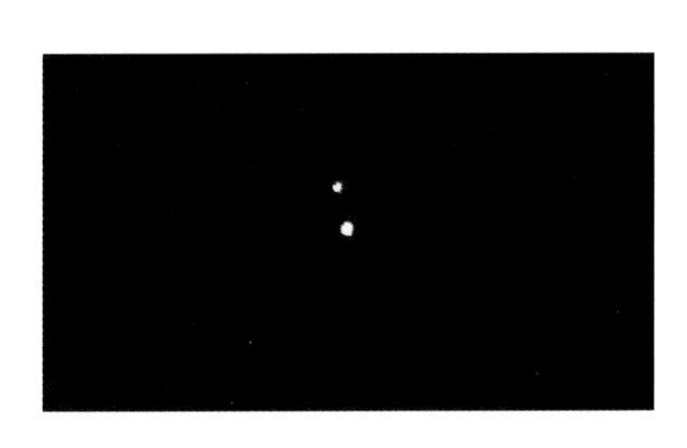

Top: Known as the Trapezium, the quadruple star at the heart of the Orion Nebula is one of the most famous and impressive multiple stars in the heavens. Bottom: The beautifully colored double star Albireo is visible in the smallest backyard telescopes.

Variable Stars

Although our Sun, being a single star, is in the minority, its light output remains stable over long periods of time. For thousands of years, the Sun has not varied from its present brightness by much more than 1 percent, and most other stars are similar

stable thermonuclear furnaces with uniform energy outputs. However, a few stars are passing through critical stages in their evolution, where their thermonuclear generators are shifting from one type of fuel to another. What happens at this stage varies from star to star, depending mostly on the star's mass, but some stars undergoing this transition can oscillate in brightness by a factor of 15,000 in a span of just one year. Astronomers are keenly interested in these variable stars. Understanding them may reveal considerable information about stellar evolution and about the onset of death among stars, since many variable stars appear to be nearing the end of their lives.

Roughly a dozen variable stars are visible to the unaided eye. Nearly 100 can be identified with binoculars, and a backyard telescope brings thousands into view. Using 10-inch and smaller telescopes, Ray Thompson of Maple, Ontario, one of the world's most experienced variable-star observers, has made more than 11,000 individual visual estimates of the brightness of these fluctuating suns over the past 40 years. He is one of several hundred amateur astronomers who observe variable stars and report their brightness estimates to the American Association of Variable Star Observers (AAVSO) in Cambridge, Massachusetts. The AAVSO, in turn, tabulates the observations and forwards the results to variable-star specialists at major observatories, who have neither the time nor the staff to keep a nightly tab on the sky's population of variables.

Apart from its opportunities for assisting in scientific research, variable-star observing is one of the best ways to increase the eye's sensitivity for distinguishing faint objects and for detecting brightness differences. Furthermore, the process of hunting for the stars significantly expands the observer's knowledge of the sky. I began observing variables in my second year of telescopic exploration. I made more than 1,000 brightness estimates in the next two years, but for me, the experience was most valuable as an apprenticeship to learning celestial geography.

A few of the brightest variable stars are labeled on the charts in this chapter. Stars of suitable comparison magnitudes located nearby are identified to one-tenth of a magnitude. To practice calculating a variable star's brightness, pick a star slightly brighter or fainter than the variable, then make an estimate of the variable's magnitude. Predictions of the maximum brightness of a few prominent variables are given in the *Observer's Handbook* (see Resources).

Two contrasting open star clusters, M46 (left) and M47, are easy binocular targets, but it takes a telescope to reveal individual stars in M46. For more information, see Chart 17 on page 116.

There are four main classes of variable stars, all of which have examples indicated on the main charts. *Cepheid* variables, named after the prototype Delta (δ) Cephei, are highly regular pulsating stars. Their period of variability and their range in brightness are so precise that they are used as "standard candles" to determine the distances to galaxies beyond the Milky Way (see box, page 89).

Eclipsing variables are binary systems in which neither star varies, but their orbits alternately place one star in front of the other, thus producing an eclipse and a decrease in light. The star Algol in Perseus is the best-known member of this class.

Long-period variables are red giants similar to Betelgeuse and Antares but at a different stage in their evolution. Some of them vary by more than 10 magnitudes in less than a year, and it is not unusual for these stars to change by one or two magnitudes in a few weeks. These are the favorites of backyard variable-star observers because of the wide magnitude range and because the cycles do not exactly repeat each other, introducing an element of anticipation to a night's observing.

The *irregular* variables are the final type of easily observed variable star and include a variety of oddball stars, some that are normally bright and then become dimmer and others that are normally faint and occasionally brighten. Another group of irregular variables ponderously oscillates over months or years, brightening or fading by a few tenths of a magnitude. Betelgeuse is in this category.

The most dramatic variable stars are *novas*, which unpredictably blast off their outer layers, the explosion causing the star's brightness to shoot up by 12 to 15 magnitudes. Normally, these are totally obscure, very faint stars. The brightness surge occurs over a period of hours or days so that in effect, a "new" star appears. The nova of 1975 in the constellation Cygnus was, at its brightest, almost equal to Deneb. Two nights later, it had dropped to third magnitude and, in a few weeks, was visible only with binoculars. Because of the sudden appearance of a nova, the backyard astronomer who is familiar with the sky can be the first to spot one. Since novas are relatively rare (only one or two a decade reach third magnitude or brighter), a nova discovery is a major astronomical event. A nova occurs in a close binary-star system in which one of the stars is a dense white dwarf whose intense gravity is vacuuming up material from its companion. Eventually, the new hydrogen fuel falling on the star becomes hot and dense enough to ignite nuclear fusion reactions. The energy released blasts off the captured star-stuff, producing a brilliance that lights up a sector of the galaxy for several weeks. The star returns to normal after a few months, and decades or centuries later, the cycle repeats itself. Only a few novas have cycles shorter than a human lifetime, and these are diligently watched by seasoned variable-star observers.

The rarest class of variable star is the *supernova*, which is the sudden explosive death of a massive star. A supernova occurs when a star's central thermonuclear furnace runs short of fuel and shuts down. The star collapses, but the heat generated by the collapse produces a ferocious fireball billions of times brighter than the Sun. In a few hours, an apparently normal star can become almost as bright as a galaxy! The supernova slowly dims during the next year or two.

Superluminous stars like Rigel are supernova candidates. If Rigel were to become a supernova, it would be hundreds of times brighter than the full Moon for weeks. But supernovas are so rare, the chances of one occurring as close as Rigel are exceedingly slim. The last known supernova in our galaxy was observed by German astronomer Johannes Kepler in 1604. It was as bright as Jupiter. In February 1987, a supernova erupted in the Large Magellanic Cloud, the Milky Way's nearest companion galaxy. Supernova 1987A, as it was called, reached third magnitude, but its location in the far southern sky made it invisible from anywhere north of Central America.

Star Clusters

In Chapter 4, we identified the Hyades and the Pleiades, in the constellation Taurus, two clusters of stars several hundred light-years away that are visible to the unaided eye. Hundreds of similar clusters are scattered around our sector of the galaxy. They range from modest collections of a few dozen suns to swarms of thousands of stars kept in a huddle by their mutual gravity. These aggregations are known as open, or galactic, star clusters, but more often, they are simply called star clusters. They offer one of the truly rewarding categories for deep-sky hunters and casual observers alike. At least 20 star clusters are easy

The Double Cluster in Perseus (see Chart 19 on page 118) ranks as one of the great deep-sky treasures of the northern sky. The two stellar jewel boxes are impressive in any binoculars or telescope.

targets for binoculars, and a handful—such as M7 in Scorpius, the Perseus Double Cluster and the Beehive cluster (M44) in Cancer—are exquisite objects in binoculars.

The Pleiades

Just under 400 light-years away is the most prominent star cluster in the sky, the Pleiades, a group that is often mistaken for the Little Dipper because of the arrangement of its six brightest stars. The Pleiades are sometimes called the seven sisters, a reference to the seven stars visible to people with slightly better-than-average eyesight. I have a tough time seeing more than six stars with the unaided eye, even under excellent conditions, although some of my astronomy students have reported seeing as many as 11. Several dozen Pleiades stars can be detected in binoculars and a hundred or more by telescope.

The approximately 400 stars in this cluster were all born about 20 million years ago, which makes them baby stars compared with the 4.7-billion-year age of the Sun. Alcyone, the brightest of the Pleiades stars, is 500 times brighter than our Sun. It shines with intense blue-white light, as do all the youthful stars in this cluster. The cluster is roughly 50 light-years in diameter. Near its center, the stars average two light-years apart, about 50 times the stellar density in our vicinity.

In his classic *Celestial Handbook*, the late astronomer and historian Robert Burnham notes a connection between the Pleiades and Devils Tower, the wonderfully impressive rock formation that rises like a colossal petrified tree stump 264 meters above the plains of northeastern Wyoming. According to Kiowa lore, Devils Tower was created by the Great Spirit to protect seven Indian maidens who were being pursued by giant bears. The maidens were afterward placed in the sky as the Pleiades cluster, and the marks of the bears' claws can still be seen in the vertical striations on the sides of Devils Tower.

Certainly, the Pleiades star cluster has prompted inquiring gazes for many centuries. Chinese records include references to this celestial group as far back as 2357 B.C. However, if we could somehow be transported back in time to the Age of Dinosaurs, the sky would not be adorned by the Pleiades. They would not be born for another 50 million years.

The Pleiades, the brightest star cluster visible from Earth, is obvious to the naked eye in late autumn and winter. The nebula that entwines the cluster, so prominent in long-exposure photographs like this, is only faintly visible by telescope under dark skies.

The Big Dipper Cluster

Studies of the motions of the seven Big Dipper stars over several decades have revealed that all but two of them are part of a star cluster. The two nonmembers are Dubhe and Alkaid, at the configuration's opposite extremities. The other five are all related. Astronomers estimate that they were born in the same region of space about 200 million years ago.

This group plus about 30 other stars scattered across the sky make up the nearest star cluster. The reason it does not look like a star cluster is that we are inside it,

Distances to Stars & Galaxies

If the distance from Earth to the Moon were reduced to the same distance that this page is from your eyes, the most distant planets, Neptune and Pluto, would be six kilometers away. But the nearest star beyond the Sun, Alpha Centauri, would be 30,000 kilometers away—almost one-tenth the *real* distance to the Moon.

The first star distance was determined in 1838 by German astronomer Friedrich Wilhelm Bessel, using a method that astronomers have employed ever since. A star's position in the sky relative to the stars that appear around it is recorded photographically. The process is repeated six months later, when Earth has swung to the other side of its orbit. If the star is within about 100 light-years, it will have shifted noticeably against the background pattern of more distant stars, because we are viewing it at a slightly different angle from the other side of the Earth's orbit. A measurement of this shift, called parallax, will give the distance to the star.

Parallax measurements using Earth-based telescopes produce reliable distances to stars within 50 light-years of Earth. Accuracy decreases rapidly after that, and by 200 light-years, educated guesses based on the star's type and age are often just as good. During the last half of the 20th century, the technique pretty well reached its limit. Ever-present turbulence in the air distorts star images and has severely limited further gain, despite improved telescopes.

Enter Hipparcos, a European Space Agency satellite designed specifically to measure the parallaxes of the brightest 100,000 stars with 10 to 100 times the accuracy that Earth-based telescopes can achieve. Hipparcos was launched in 1988 and operated for six years. It took four years to analyze the satellite's readings, which were finally released in 1998. Almost every naked-eye star had its distance significantly upgraded, a boon to researchers and a nice gift to stargazers.

The charts at the end of this chapter present the Hipparcos distances to more than 250 stars—almost every star visible in mid-northern latitudes down to magnitude 3.5 and a few which are fainter than that. These distances are generally accurate to within one light-year at a distance of 40 light-years and to within 10 light-years at 400 light-years. After that, the Hipparcos accuracy drops rather swiftly until around 1,000 light-years, at which point the measurements become only marginally more reliable than educated guesses. Keep this in mind when you refer to the distances marked on the charts.

The vast majority of stars in our galaxy, not to mention other galaxies, are too remote for parallax measurements (they are part of the background that allows the parallax principle to work). Astronomers must turn to estimates rather than direct measurements. The estimates are based on the idea that stars with the same temperature and spectrum likely have the same intrinsic brightness, called absolute magnitude—like two light bulbs with the same wattage. Since we know the absolute magnitudes of the nearby stars whose distances can be measured, a star which is suspected of having the same absolute magnitude but which *appears* dimmer should be proportionally farther away.

Here's an analogy: Imagine looking at a city's streetlights from a hill. The more distant lights appear dimmer, even though they may be the same wattage as nearby lamps. Assuming that you know the wattage (absolute magnitude) of the lamps and the distance to the nearest ones, you could figure out the distances to all of them by measuring their apparent brightness with a sensitive light meter. The light meters astronomers use on their telescopes for measuring star brightness are called photometers.

The next step is the distance to other galaxies. To gauge these distances, astronomers observe pulsating stars called Cepheid variables. Cepheid variables are crucial distance calibrators because their intrinsic brightness is known to be related to their rhythmic oscillations in luminosity. For instance, a Cepheid with a long period of oscillation has high intrinsic brightness, like a 100-watt light bulb, whereas a short-period Cepheid would be a 40-watt bulb, and so on.

Fortunately, Cepheids are typically more than 1,000 times brighter than the Sun and can be seen across vast distances. They have been used as distance calibrators out to 60 million light-years, which includes several dozen nearby galaxies.

The distances to more remote galaxies, all the way to the edge of the visible universe, are estimated from their redshifts, that is, their velocity of recession from us caused by the expansion of the universe. This method is open to different interpretations. One team of astronomers might calculate a galaxy's distance from us as three billion light-years, while another would assign a five-billion-light-year figure. Such wide discrepancies have been debated for three decades. Astronomers hope that the Hubble Space Telescope and a new generation of giant Earth-based telescopes will resolve the discordant distance estimates for galaxies beyond a few million light-years from Earth.

The eighth-magnitude star cluster NGC6939 and the ninth-magnitude spiral galaxy NGC6946 offer a dramatic size and distance comparison. The cluster consists of a few hundred stars about 6,000 light-years away, located within the Perseus Arm of the Milky Way. The galaxy contains at least 50 billion stars and is 55 million light-years distant. Yet the two objects appear roughly the same brightness in backyard telescopes.

although the Sun is not a member. The Big Dipper cluster, or the Ursa Major cluster, as it is properly called, is slowly overtaking the Sun, much like a group of joggers running in a clump gradually overtaking and passing a slightly slower runner. Millions of years in the future, the cluster will be an inconspicuous patch in the direction of the star Deneb.

Our descendants will witness the gradual breakup of the Big Dipper. The five cluster stars are moving in formation in the direction of the Dipper's handle, while Dubhe and Alkaid are drifting in the opposite direction. This will bend the handle down at the end and open the bowl, but it will take many thousands of years. Our great-great-grandchildren will still see the same Big Dipper that we know today.

As star clusters go, the Big Dipper cluster is a loose aggregation. If the stars were once bound by mutual gravitation into a compact group like the Pleiades, the bonds have since loosened. Today, they are just floating in space beside each other. In a few hundred million years, they will be completely dispersed, just as the cluster from which our Sun formed lost its identity long ago.

Nebulas

Drifting among the stars in the galaxy's spiral arms are thousands of vast clouds of gas and dust called nebulas. Most of them are dark and invisible, sometimes producing the dark rifts and patches that give the Milky Way its ragged, segmented appearance. These clouds are the largest objects in the galaxy—massive celestial smog banks that block the light of millions of stars. Generally, their presence is revealed only by a paucity of stars along certain sectors of the Milky Way. Occasionally, though, they produce spectacular ethereal vistas when stars illuminate the normally dark veils of gas and dust.

Nebulas are the galaxy's maternity wards, where new stars are being born. The process begins inside the nebula, where gas and dust collect in knots, then disperse, only to collect elsewhere. Sometimes, the cloud's equilibrium experiences a major disturbance (perhaps due to a nearby supernova blast or a merger of two or more clouds), inducing pockets of cloud to collapse into denser clumps. Once a rapid infall of matter begins, a gravitational chain reaction is triggered. Atoms and molecules bang into each other and into the cloud's dust grains. Energized by such collisions, the dust radiates heat. But the opaque cloud traps the radiation, heating it further. Meanwhile, matter is piling up at the gravitational core of the pocket, and the temperature soars. The core mass attracts more matter, and the cycle escalates. In tens of thousands of years—a brief span of time by astronomical standards—temperatures climb from minus 250 degrees C to 15 million degrees, the ignition point for fusion reactions. A star is born.

The details of this process are still unclear, because the infalling envelope of gas and dust hides the stellar birth from view. The infant star, called a protostar, nestles in the womb of the mother cloud. Eventually, like afterbirth, the star rids itself of its cloak by the sheer force of its radiation. A section of the cloud is blown off, the youthful star exposed. The star will likely not be alone. The process that produces one usually acts on a substantial zone of the giant cloud, and a star cluster emerges as if bursting from a cocoon. This is exactly what is seen in several sectors of the Milky Way. The best-known and brightest of these star-birth regions is the Orion Nebula, just below Orion's belt.

The Orion Nebula

At the core of the Orion Nebula is a cluster of stars born just 50,000 years ago, according to current estimates. These are energetic suns, far more massive and luminous than our own. They have swept back the dust veils, illuminating an awesome 20-light-year-wide bowl-shaped cavern—the visible zone that astronomers call the Orion Nebula—on the edge of a titanic dark cloud that fills most of the constellation Orion.

More of the dark cloud will become exposed in the millennia ahead. Studies with the Hubble Space Telescope confirm that new stars are forming in the thick gas and dust immediately behind the Orion Nebula. As the new stars evolve and emit more heat and light, they will evaporate the dust and ionize the gas,

After the Orion Nebula, the Lagoon Nebula in Sagittarius, left, is the brightest star-forming region visible to observers in the northern hemisphere. The nebula's telescopic appearance is shown on page 96.

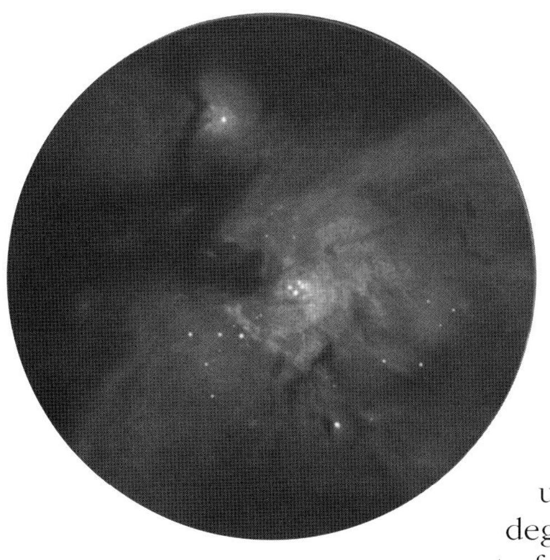

causing it to glow—just as the stars at the core of the Orion Nebula have done. Star birth is like a spreading infection: once it gets started, the process eats away at the giant interstellar cloud.

The Orion Nebula is the nearest bright nebula, at 1,400 light-years away, and is the only one plainly visible to the unaided eye (from north of +20 degrees latitude). It looks like an out-of-focus star, a fourth-magnitude puff of cosmic cotton. Binoculars reveal a delicate, hazy cup-shaped patch of light. The soft glow offers a captivating contrast to the three second-magnitude stars in Orion's belt and the third- and fourth-magnitude stars closer to the nebula.

Binoculars also disclose a fifth-magnitude star near the nebula's center. This is Theta-1 Orionis, one of the most intriguing star systems in the sky. Appearances are not deceiving here: Theta-1 *is* actually at the heart of the Orion Nebula and is the main source of its illumination. Even a 60mm refractor will reveal that this is not a single star but four arranged in a trapezoid. Called the Trapezium, the four range in brightness from fifth to seventh magnitude and are far enough apart (the smallest separation is nine seconds of arc) to make an exquisite scene—four blue jewels embedded in a delicate celestial cloud.

Photographs of the Orion Nebula capture the vast extent of the wisps and veils that cover an area almost a degree across. Yet an initial look at the nebula through a telescope may cause the observer to wonder whether this really is the same object that the photographs show. Why is it so faint? And where is the color?

When you use the lowest-power eyepiece, the nebula—at first—looks like a small, moderately bright patch around the Trapezium. But if there is no Moon and the sky is free from haze or artificial light, a glorious sight will unfold. As the eye becomes accustomed to the low light levels, subtle rifts of the nebula appear, just at the threshold of vision. The larger the tele-

The illustrations on this page emphatically demonstrate the difference between the photographic and the visual appearance of a deep-sky object such as the Orion Nebula. Right: The great nebula blazes in glorious color below the equally dazzling Horsehead Nebula. In contrast, the view of the Orion Nebula in 11x80 binoculars, above, is colorless and subtle. Top: A medium-sized telescope adds significant detail. What the sketches cannot show is the ethereal delicacy of the visual scenes.

scope, the more that can be seen. But even a small instrument will give delightful views of the distant star factory.

The human eye, however, has limitations. Its color sensitivity at low light levels is practically nil. To me, the nebula has never appeared anything but pale gray-green, even in a 16-inch telescope. Dark-adapted eyes can detect ripples, loops and just a hint of "texture" in the nebula's brighter sections. Try higher powers on these regions; this is one of the few nebulas that do not become washed out at high power. In larger telescopes, the nebula's core is a magnificent sight. The Orion Nebula's cloud-like appearance in photographs is somewhat deceptive. One can almost imagine winging through it in a spacecraft, moving among its wisps and swirls like an airliner passing through cumulus clouds. The actual situation is very different: a hypothetical spaceship rocketing through the nebula would encounter only slightly more particles than those recorded in interstellar space. The average density of the nebula is one-millionth the density of a good laboratory vacuum. The density increases near the sites where new stars are just beginning their lives or where nascent stars are being nurtured. Yet the nebula is so vast—more than 2,000 cubic light-years—that it contains enough material to form hundreds of new suns.

Averted Vision

When observing the Orion Nebula or any object of low surface brightness, experienced deep-sky observers use a technique known as averted vision. The concept of averted vision is to concentrate on the celestial object without looking directly at it. With the object of interest centered in the eyepiece field, direct your gaze toward the field's edge. Often, details invisible with direct vision suddenly become apparent. It works because the eye's visual receptors away from the central axis of vision are more sensitive to dim light. This low-light peripheral vision, developed in human eyesight millennia ago (our ancestors must have needed it for self-preservation), should always be employed when observing objects at the threshold of vision—either stars or nebulas—with or without a telescope.

Always start your examination of nebulas at low power, then work up. In many instances, the lowest power offers the best view, because the object appears small but brighter than at higher magnification.

Above: M13, the Hercules cluster, is the favorite globular cluster for many backyard astronomers because of its optimum position in the night sky—it passes almost directly overhead during late-spring and early-summer evenings (see Chart 6 on page 105).

These two illustrations correspond to the cluster's visual appearance in 8- and 15-inch telescopes. Aperture really does count with globular clusters. However, the same effect is partially achieved through a technique known as averted vision.

Globular Clusters

More distant than the galaxy's open star clusters and nebulas are globular star clusters, which are, in effect, tiny satellites of the Milky Way Galaxy. At least 140 globular clusters surround the galaxy, about one-third of which are visible in backyard telescopes. Omega Centauri, at fourth magnitude, is the brightest of these and is easily visible to the unaided eye. Unfortunately, because of its location in the southern sky, it is visible only from the extreme southern United States or farther south. The brightest globulars visible from midnorthern latitudes are M13 and M22, in Hercules and Sagittarius, respectively, which are both easy binocular targets.

These great spherical swarms of stars range from 50 to 200 light-years in diameter and have populations of up to two million suns. Backyard telescopes less than 4 inches in aperture will show them as concentrated balls of light gradually fading off at the edges, while larger instruments will resolve some of the brightest stars. Averted vision improves the view substantially. More than any other class of celestial object, globular clusters benefit from increases in telescope aperture. In large telescopes, these clusters are stunningly dramatic sights.

Galaxies

Everything described so far is part of the Milky Way Galaxy. The final category of deep-sky objects is other galaxies—distant islands of billions of stars that float in the cosmic emptiness out to the greatest distances that telescopes can penetrate.

There are two classes of galaxies available to backyard astronomers: spiral and elliptical. Spiral galaxies are similar to the Milky Way. Some galaxies have their twirling arms tightly wound, producing almost uniform disks, while others have loose, ragged configurations. Spiral galaxies come in a limited range of sizes, from systems containing a few tens of billions of stars to giants with several trillion suns. Many spiral galaxies are approximately the size of the Milky Way, which appears to be about average.

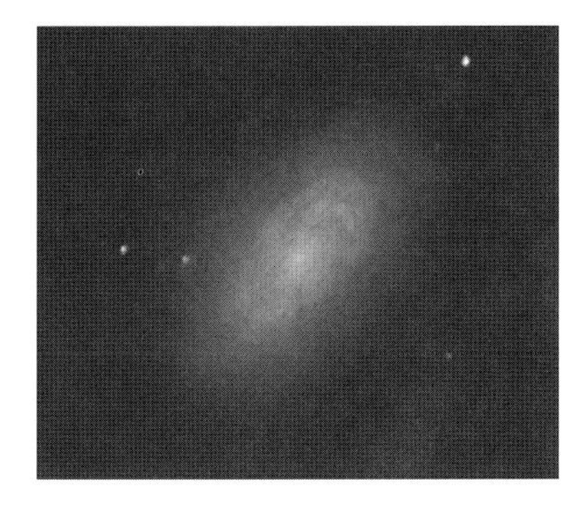

Elliptical galaxies, basically featureless spherical systems of stars, have a far greater range of masses, from dwarf ellipticals, with a few million stars, to the colossal supergiant elliptical galaxies that boast up to 100 trillion stars. These are the largest objects that backyard telescopes can detect. Some can be seen at distances of 100 million light-years or more.

As with nebulas, the details that long-exposure photographs reveal in galaxies can never be seen with the eye. However, the shape of a spiral galaxy is often evident. In general, a bright nucleus that appears like a star embedded in a small patch of mist is surrounded by a fainter envelope of light. Occasionally, there is a hint of spiral arms, but more often, the spiral galaxy will simply be a general haze around the brighter nucleus. Edge-on spirals look like delicate slivers of light. The Milky Way's two companion galaxies, the Large and Small Magellanic Clouds, are deep in the southern sky and are not visible from anywhere in Canada or the United States.

The nearest galaxy similar to our own is the Andromeda Galaxy, seen as a hazy fourth-magnitude object near the star Nu (ν) Andromedae. At 2.4 million light-years, the Andromeda Galaxy is the most distant object that the unaided eye can detect. Binoculars reveal its oval outline, three to four degrees long by less than one degree wide. Because the galaxy is tipped only 18 degrees to our line of sight, it has a distinct oval shape. The galaxy's two small elliptical companions, M32 and NGC205, are visible in binoculars, but not easily.

Beginners are often profoundly disappointed by their first telescopic view of the Andromeda Galaxy. It looks so magnificent in photographs. Even lowly binoculars offer a more pleasing view than can be seen in a telescope. At least substantial structure is seen in the Orion Nebula, which also produces spectacular photographic portraits. What's going on here?

The Andromeda Galaxy is a *big* object. A wide field of view —wider than many beginners' telescopes can offer—is essential. The best views are with a pair of oversized binoculars or a telescope with a 2-inch focuser and a 30mm to 40mm 2-inch eye-

Galaxies are elusive telescopic targets for the recreational astronomer. Most are small and faint, mere oval smudges, with only hints of spiral arms or other visible detail. The three sketches on this page (left, above and top right) are typical of what can be glimpsed in an 8-to-12-inch telescope—unimpressive on the one hand but astounding when you realize that each smudge is a galaxy similar to the Milky Way yet 40 to 60 million light-years distant. The color images of galaxies M51, facing page, and NGC6946, top left, were taken with a 25-inch telescope.

Star Diameters

Since stars are too remote for any telescope to measure their diameters directly, how do astronomers know that Sirius is 1.8 times the diameter of our Sun or that Aldebaran is 45 times wider? The answer is *indirect* measuring techniques, which have yielded the diameters of several hundred stars.

The first method involves the precise electronic monitoring of a star when it is blocked out by the Moon. As the Moon makes its monthly orbital trek around Earth, it passes in front of many stars, but only occasionally are the stars bright enough for the detection equipment to complete the experiment. Although the technique is little more complicated than observing the length of time it takes for the star to disappear (which is almost instantaneous), it has revealed accurate diameters for several dozen stars.

However, the Moon's path is limited to a specific sector of the sky, and astronomers anxious to determine the sizes of other stars developed two exceedingly painstaking techniques—stellar interferometry and speckle interferometry—that ultimately required years of refinement to capture the diameters of a handful of stars. Both methods involve electronic analyses to cancel out the interference of the Earth's atmosphere and to take advantage of the physical properties of light propagation. To appreciate the difficult nature of these experiments, consider the fact that a typical star seen from Earth is about the same size as a walnut on the Empire State Building viewed from a distance of 600 kilometers, or as far away as Toronto.

The techniques of stellar interferometry, speckle interferometry and electronic monitoring of lunar occultations have produced the diameters of some of the stars mentioned in this book. The diameter of a star can also be inferred from stellar-evolution theory, based on the calibrations provided by the diameters of stars that have been measured. In this way, almost any star can now be tagged with an estimated diameter.

In tallying up the diameters, we find that among all the naked-eye stars, less than 1 percent are smaller than our Sun. And yet a census of *all* stars shows that the average star is both smaller and dimmer than the Sun. Thus the incongruity of the starry sky as seen from the backyard: It is the giants, the blazing beacons of the galaxy, that make up the familiar constellations. The average citizens of our starry vault make a negligible contribution to the night sky.

This illustration shows the baked surface of a hypothetical planet of the star Betelgeuse. At 800 times the Sun's diameter, this is the largest star within 1,000 light-years of our solar system.

piece, which places the galaxy in a two-degree or wider field.

For more remote galaxies, observers will be content with just seeing these giant cities of stars, their pale, delicate forms stimulating the mind more than the eye. For example, consider M51, the Whirlpool Galaxy, which is located about one-fifth of the way from Alkaid, in the Big Dipper, to Cor Caroli, in the small constellation Canes Venatici. Although equal in size to the Milky Way Galaxy, M51 is 35 million light-years away and is merely a faint smudge in small telescopes. Telescopes over 4 inches in aperture may show it as a double smudge.

The Whirlpool Galaxy is one of the few galaxies whose spiral structure can be glimpsed with typical backyard-astronomy equipment. I have never seen it distinctly with anything less than a 10-inch, although I was once able to detect two spiral arms using a 5-inch, but it was a threshold observation. The true magnificence of this continent of stars, however, can be seen only in long-exposure photographs, which also reveal that the "double smudge" effect is due to a companion galaxy apparently dangling at the end of one of M51's spiral arms. The companion, known as NGC5195, is another spiral galaxy behind M51 that was severely distorted a few hundred million years ago during a close encounter with M51. This galactic sideswipe wrecked the shape of NGC5195 and warped M51's symmetrical spiral.

Such near collisions are not uncommon in the realm of galaxies. What is unseen, however, even in photographs, is the billions of stars that must have been wrenched away from both of these galaxies and flung into the abyss of intergalactic space. If our galaxy suffered a similar experience and our Sun were torn from its parent galaxy, Earth would remain unaffected, continuing in its solar orbit while the Sun drifted forever in the intergalactic darkness. The sky would be almost totally black, punctuated here and there by hazy islands—the nearest of the universe's billions of galactic star cities.

Telescope Experience

It takes a long time to get used to looking through a telescope, especially when the target is a dim deep-sky object. Some of my astronomy students are unable to see anything when they step up for their first look. A big problem seems to be keeping the head steady while bending over to look in the eyepiece. Observing with a telescope means training the body as well as the mind. It requires that the observer stand rigidly still but relaxed. The eye and the mind must slowly be trained to pick out details that are initially imperceptible. The telescope's lowest magnification should always be used at first. This makes it easier to locate, focus and see the celestial target clearly.

The delicate spindle shape of an edge-on spiral galaxy or the filamentary tendrils of a distant cloud of dust and gas like the Orion Nebula are sights that are almost invisible to the novice. After several weeks, my students are usually able to see details that went completely undetected when they initially looked

Designation of Sky Objects

The all-sky charts in Chapter 4 provide the names of the constellations and a few of the brighter stars. The charts in this chapter show many more stars, as well as the locations of a variety of deep-sky objects. Since only a hundred or so of the brightest stars have been assigned individual names (Vega, Arcturus, and so on), there are other naming systems to identify the less prominent stars.

The oldest systematic attempt at star naming was developed by German astronomer Johann Bayer at the beginning of the 17th century, just before the invention of the telescope. Bayer designated the stars in each constellation in order of brightness by using lowercase letters of the Greek alphabet: alpha (α) for the brightest, beta (β) for second brightest, gamma (γ) for third brightest, and so on. For some of the larger constellations, Bayer used the entire 24-letter Greek alphabet.

By the end of the 17th century, astronomers realized that they had to remedy the limitation of Bayer's system. British astronomer John Flamsteed suggested assigning a number to each star in a constellation, thus eliminating the confining Greek alphabet. He applied numbers from the western side of a constellation toward the east, including each star within the limit of naked-eye visibility. The largest constellations received more than 100 of Flamsteed's numbers. But the Flamsteed system did not supersede the Bayer designations, and today, Flamsteed numbers are used only on the stars not covered by the Greek letters. Stars too faint for the Flamsteed list are each identified by one of several more recent catalog numbers generated at observatories specializing in star positions. For example: BD36°2516, from the German *Bonner Durchmusterung* star catalog.

Deep-sky objects have their own identification system, usually derived from the catalog of 18th-century astronomer Charles Messier or the *New General Catalogue* compiled by English astronomer J.L.E. Dreyer and published over a century ago. The 109 objects in the *Messier Catalogue* are designated M1, M2, and so on. The several thousand *New General Catalogue* objects are prefixed by the letters NGC. Most of the Messier objects are in the *New General Catalogue*, many of them having a popular name as well. M1, for example, is known as NGC1952 and also as the Crab Nebula. A few objects not in either of these major catalogs have different prefixes, such as IC or Col., identifying other lists.

Star atlases have their own language, but once you become accustomed to it, a wealth of information is available. Above: A small section of the revised (1998) edition of *Sky Atlas 2000*, which shows stars to magnitude 8. This and other recommended atlases are described in Chapter 12. Top: Star clusters M35 (large) and NGC2158 in Gemini. Right: M16, the Eagle Nebula.

The Greek Alphabet

α	alpha	AL-fuh
β	beta	BAY-tuh
γ	gamma	GAM-uh
δ	delta	DELL-tuh
ε	epsilon	EPP-sill-on
ζ	zeta	ZAY-tuh
η	eta	AY-tuh
θ	theta	THAY-tuh
ι	iota	i-OH-tuh
κ	kappa	CAP-uh
λ	lambda	LAM-duh
μ	mu	mew
ν	nu	noo
ξ	xi	zeye
ο	omicron	OM-ih-krawn
π	pi	pie
ρ	rho	row
σ	sigma	SIG-muh
τ	tau	taw
υ	upsilon	UP-sih-lon
φ	phi	fie
χ	chi	kie
ψ	psi	sigh
ω	omega	oh-ME-guh

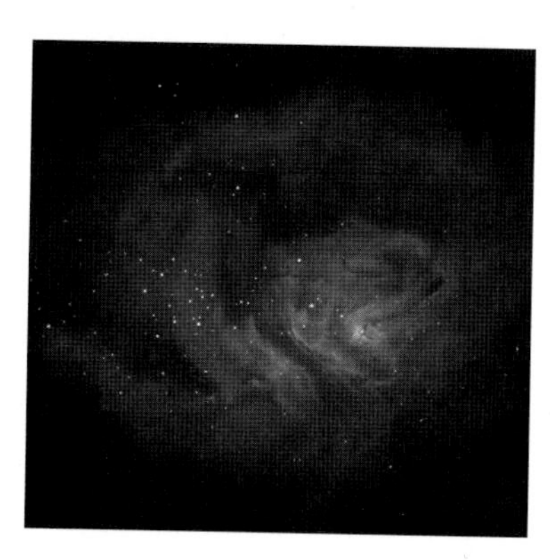

through the telescope. But they are awed by the fact that I can aim the telescope at an apparently vacant part of the sky, twiddle a few knobs and present to them a star cluster or a nebula. This is really not a special talent. All veteran backyard stargazers can do it. But it is largely a self-taught skill.

Being able to locate objects in the sky is primarily an exercise in self-edification. Someone can point out the main constellations, but when it comes to seeking the fainter objects buried among the myriad stars, every backyard astronomer must serve an apprenticeship of self-taught sky knowledge. Once the geometrical relationships of the constellations begin to link in your mind, the celestial clockwork—the sky motions caused by the Earth's rotation on its axis and by its orbit around the Sun—will soon become familiar. After a year or two, it all starts to fall into place, and the night sky becomes more than just a pretty tapestry of stars.

Eventually, most amateur astronomers come to know by memory the relative positions of the brightest 500 to 1,000 stars. That provides a celestial web from which the more interesting telescopic quarry can be tracked. This sky-familiarization process is where many beginners give up. They pack up their telescopes because they are unable to find the celestial sights. But they only shortchange themselves by limiting the depth of the hobby and underestimating their own abilities.

Others, expecting to see Technicolor vistas like the observatory photographs displayed in coffee-table astronomy books, are disappointed when confronted with the real objects observed through backyard telescopes. But it is seldom pointed out in those books that if one could look through the telescope which took the photographs, the objects would appear much less impressive. Photographic film accumulates light from a celestial source for minutes or hours, whereas the human eye generally forms a new image in the brain every one-fifth of a second. Nevertheless, a look at the subdued, but real image of a remote galaxy or nebula has a chilling effect that must be experienced to be appreciated.

Above: Comparison of a photograph (left) of M8, the Lagoon Nebula, taken through a 6-inch refractor and a sketch of the same object as seen in a 15-inch Newtonian. A small telescope can record more photographic detail in a faint object than the eye can discern using a much larger telescope. But what the photograph fails to record is the ethereal real-time impressions that the eye-brain combination gathers at the telescope eyepiece. Right: Binocular star clusters IC4756 (left) and NGC6633.

Keeping Records

No formalities or standard formats are needed for recording a night under the stars. Any type of notebook will do, although I prefer the spiral-bound high school workbooks with blank pages (for quick sketches) facing lined pages (for notes). The important point is to begin keeping notes from the first night that something is recognized. Record the date, time, place, instruments used (if any) and objects seen. If something was searched for but not found, note that too.

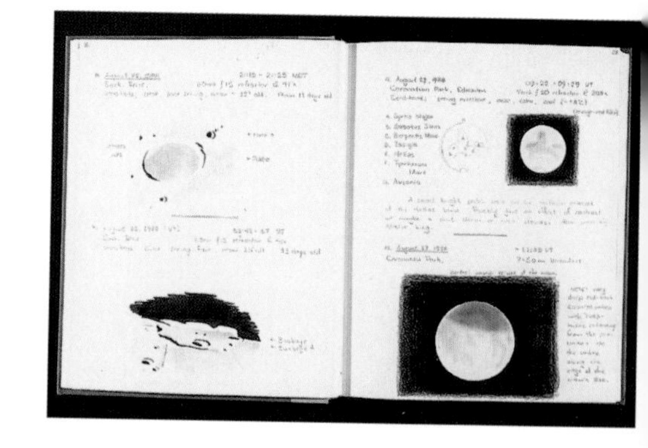

Observing conditions at your site influence what can and cannot be viewed on a given night. Hazy skies work against detection of fainter objects but do not significantly affect observation of lunar craters, bright planets or double stars. To grade the clarity of the night sky, note the faintest star seen with the naked eye near the Little Dipper (Ursa Minor). Stars to sixth magnitude are marked for this purpose on Chart 20. This provides a standard basis for comparison from one night to the next. After you have logged a few hundred objects, the book becomes a personal record of ever deeper penetration of the cosmos.

Using the Deep-Sky Charts

The 20 charts on the following pages present in detail almost the entire sky visible from midnorthern latitudes. Beyond their function as a star atlas, the charts provide information about hundreds of naked-eye, binocular and telescopic objects. Each chart includes one or two key constellations in an area of the sky about 45 by 55 degrees—roughly the size of a page in this book held at arm's length. All stars to fifth magnitude are shown. The Bayer designations of brighter stars are given, along with magnitudes to one decimal place. Other information, such as star diameter, luminosity and distance, should be viewed as increasingly less accurate the more remote the object. The reliability of deep-sky data depends primarily on the distance determinations. Distance estimates are generally very accurate out to 400 light-years, fairly good to 1,000 light-years and subject to chains of educated guesses after that. Information is color-coded.

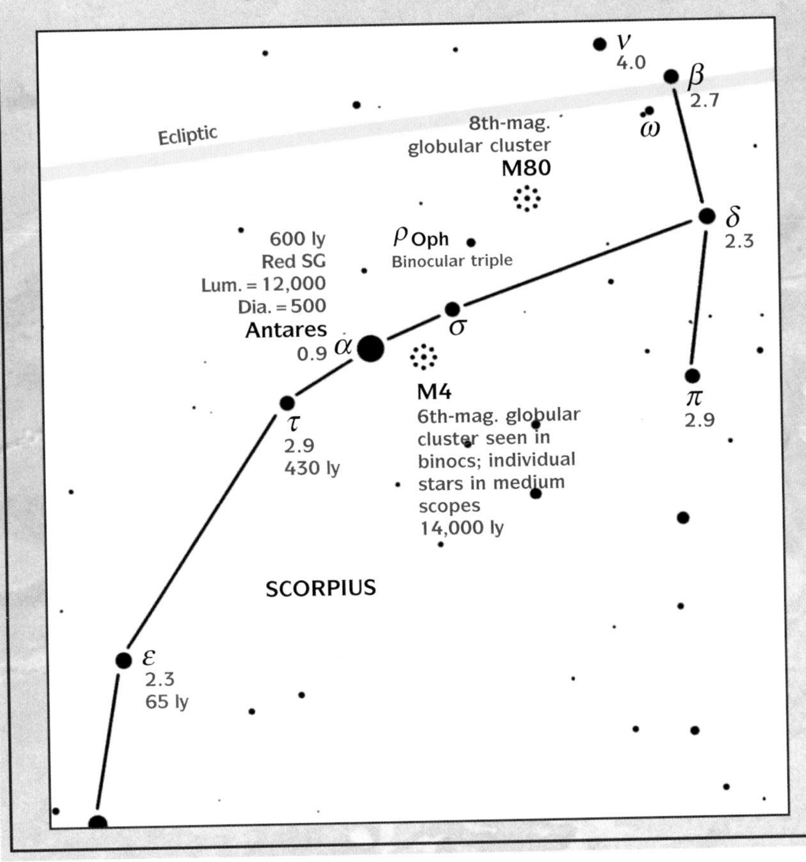

Black = **Names:** stars and constellations; deep-sky objects
Blue = **Observing information:** type or class of object; magnitude; double- and multiple-star data; general appearance; instrumentation needed
Red = **Descriptive astrophysical information:** distance; actual size; luminosity; classification

When you are using a red-filtered flashlight outdoors, the red lettering becomes less prominent. This is intentional, because this information is not essential during outdoor observing. The blue lettering—the observing information—appears black in red-filtered light.

Abbreviations, code blue

number with decimal (e.g., 0.9): visual magnitude
A=2.3; B=4.0: magnitudes of components of double or multiple star
Sep.=44": separation of components of double star in seconds (") or minutes (') as indicated
Dia.=20': apparent diameter of planetary nebula or star cluster in seconds (") or minutes (') as indicated

Abbreviations, code red

ly = distance in light-years
Lum. = luminosity compared with our Sun
Dia. = diameter of star in solar diameters
Sep. = actual separation of components of a double star
Blue SG = blue supergiant; most luminous type of star; these massive high-temperature suns burn so furiously that they last only a few million years
Red SG = red supergiant; largest type of star; believed to be a late-life transition stage of blue supergiants; the final stage may be a supernova explosion
Blue G = blue giant, a less massive, less luminous, longer-lived version of the blue supergiant
Yellow G = yellow giant, a former blue giant that is evolving toward a red-giant stage
Red G = red giant

The 20 main charts in this chapter are designed for ease of use outdoors. Each chart has essential observing information positioned directly beside the target object. A section of Chart 8 is shown above. Right: Photograph of a small section of the sky region covered by Chart 8 shows the globular cluster M4, located near Antares and Sigma (σ) Scorpii.

CHECKLIST OF DEEP-SKY TREASURES

Here are the most prominent and impressive deep-sky objects visible from midnorthern latitudes. As you work through the list, write in your first impressions.

DESIGNATION AND DESCRIPTION	CHART #	BINOCULARS	TELESCOPE	NOTES
M81 & M82, brightest pair of galaxies visible in a single telescopic field	1	•	•	
Mizar, famous double star for small telescopes	1		•	
M51, Whirlpool Galaxy, one of the brightest galaxies	2	•	•	
Algieba, fine double star for small telescopes	3		•	
M13, Hercules globular cluster	6	•	•	
M22, bright globular cluster	8	•	•	
M4, easy-to-find globular cluster beside Antares	8	•	•	
M8, Lagoon Nebula, brightest after Orion Nebula	8	•	•	
M7, spectacular cluster for binoculars	8	•		
M27, Dumbbell Nebula, brightest planetary nebula	9	•	•	
M11, rich star cluster, especially impressive in medium-sized telescopes	9	•	•	
Albireo, a favorite small-telescope double star	10		•	
Epsilon (ε) Lyrae, the famous double-double	10		•	
M57, Ring Nebula	10		•	
M31, Andromeda Galaxy, brightest spiral galaxy	13	•	•	
M45, Pleiades star cluster, the sky's brightest	15	•	•	
M42, Orion Nebula, visible to the naked eye	16	•	•	
Theta-1 (θ) Orionis, Trapezium, quadruple star at core of Orion Nebula	16		•	
M46 & M47, impressive pair of open clusters	17	•	•	
M41, easy-to-find open cluster below Sirius	17	•	•	
M44, Beehive, a naked-eye open cluster	18	•		
Double Cluster in Perseus, magnificent	19	•	•	

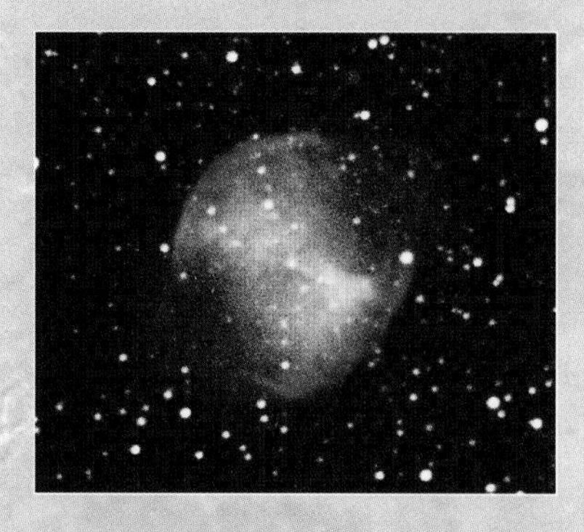

A gallery of deep-sky treasures, from top to bottom: sketch of galaxies M82 and M81 as seen in a 15-inch Newtonian; M51, the Whirlpool Galaxy; M27, the Dumbbell Nebula; M22, globular cluster in Sagittarius.

Spring
Summer
Index for Atlas Charts
Autumn
Winter
1
2
3
4
5
10
6
4
9
7
8
20
13
12
14
11
19
18
15
17
16

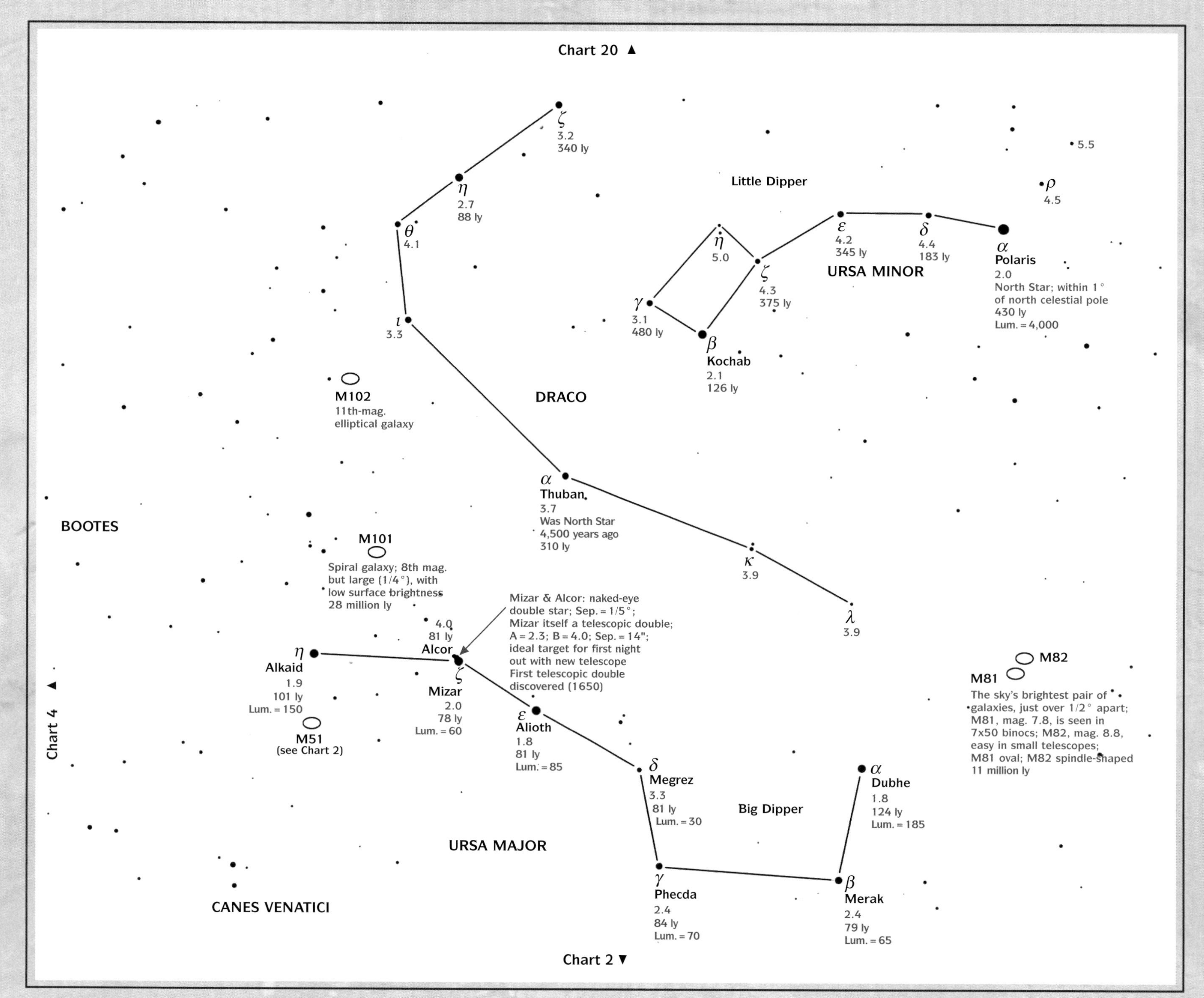

Chart 1

Ursa Major   Ursa Minor   Draco

Visible all year in north; best orientation in late winter and spring.

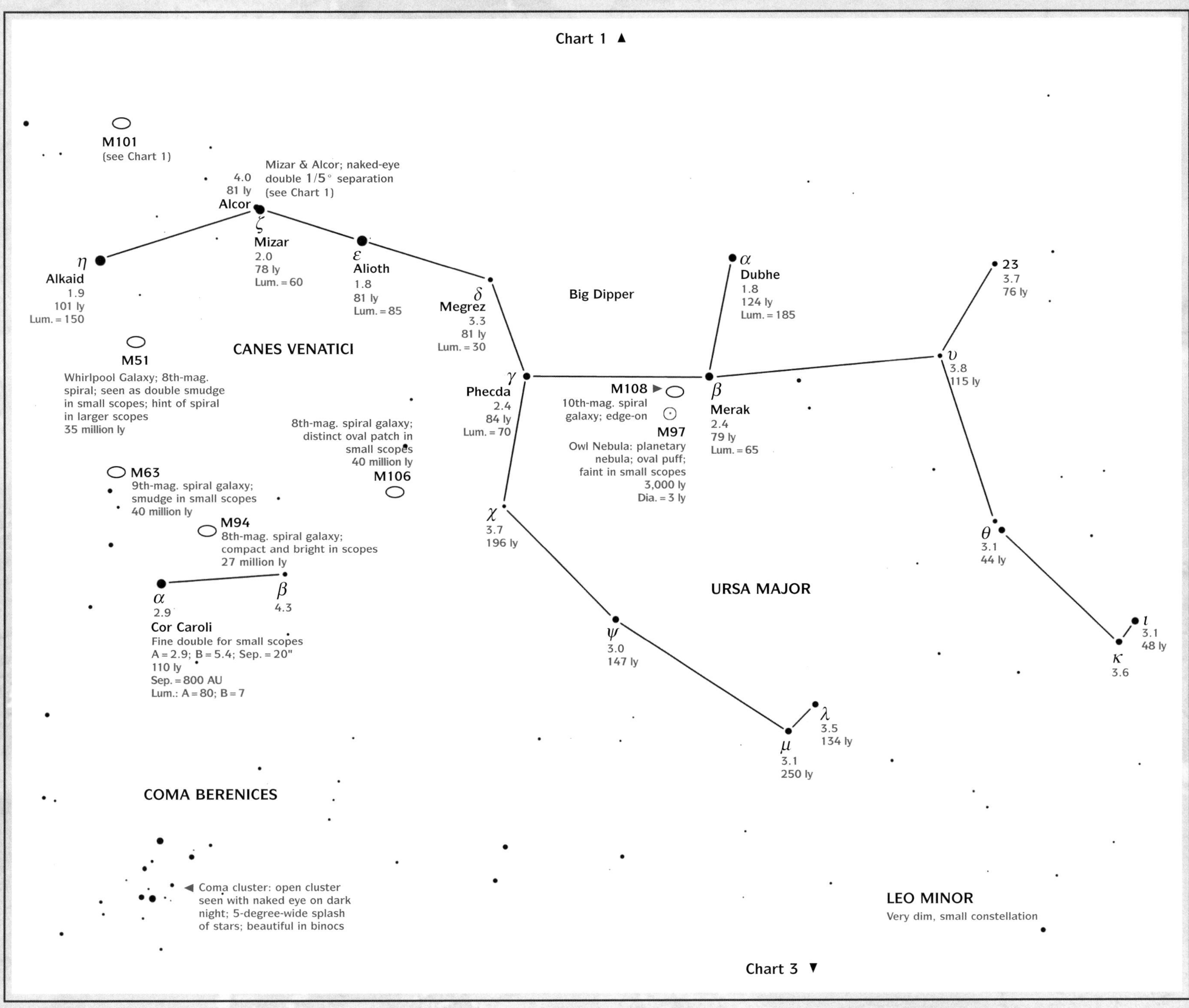

Chart 1 ▲
M101
(see Chart 1)
Mizar & Alcor; naked-eye double 1/5° separation (see Chart 1)
4.0
81 ly
Alcor
ζ
Mizar
2.0
78 ly
Lum. = 60
η
Alkaid
1.9
101 ly
Lum. = 150
ε
Alioth
1.8
81 ly
Lum. = 85
δ
Megrez
3.3
81 ly
Lum. = 30
Big Dipper
α
Dubhe
1.8
124 ly
Lum. = 185
23
3.7
76 ly
M51
Whirlpool Galaxy; 8th-mag. spiral; seen as double smudge in small scopes; hint of spiral in larger scopes
35 million ly
CANES VENATICI
γ
Phecda
2.4
84 ly
Lum. = 70
M108 ▶
10th-mag. spiral galaxy; edge-on
M97
Owl Nebula: planetary nebula; oval puff; faint in small scopes
3,000 ly
Dia. = 3 ly
β
Merak
2.4
79 ly
Lum. = 65
υ
3.8
115 ly
8th-mag. spiral galaxy; distinct oval patch in small scopes
40 million ly
M106
M63
9th-mag. spiral galaxy; smudge in small scopes
40 million ly
M94
8th-mag. spiral galaxy; compact and bright in scopes
27 million ly
χ
3.7
196 ly
θ
3.1
44 ly
α
2.9
Cor Caroli
Fine double for small scopes
A = 2.9; B = 5.4; Sep. = 20"
110 ly
Sep. = 800 AU
Lum.: A = 80; B = 7
β
4.3
URSA MAJOR
ψ
3.0
147 ly
ι
3.1
48 ly
κ
3.6
λ
3.5
134 ly
μ
3.1
250 ly
COMA BERENICES
◀ Coma cluster: open cluster seen with naked eye on dark night; 5-degree-wide splash of stars; beautiful in binocs
LEO MINOR
Very dim, small constellation
Chart 3 ▼

Chart 2

Ursa Major   Canes Venatici   Coma Berenices

Well positioned for viewing in late winter, spring and early summer.

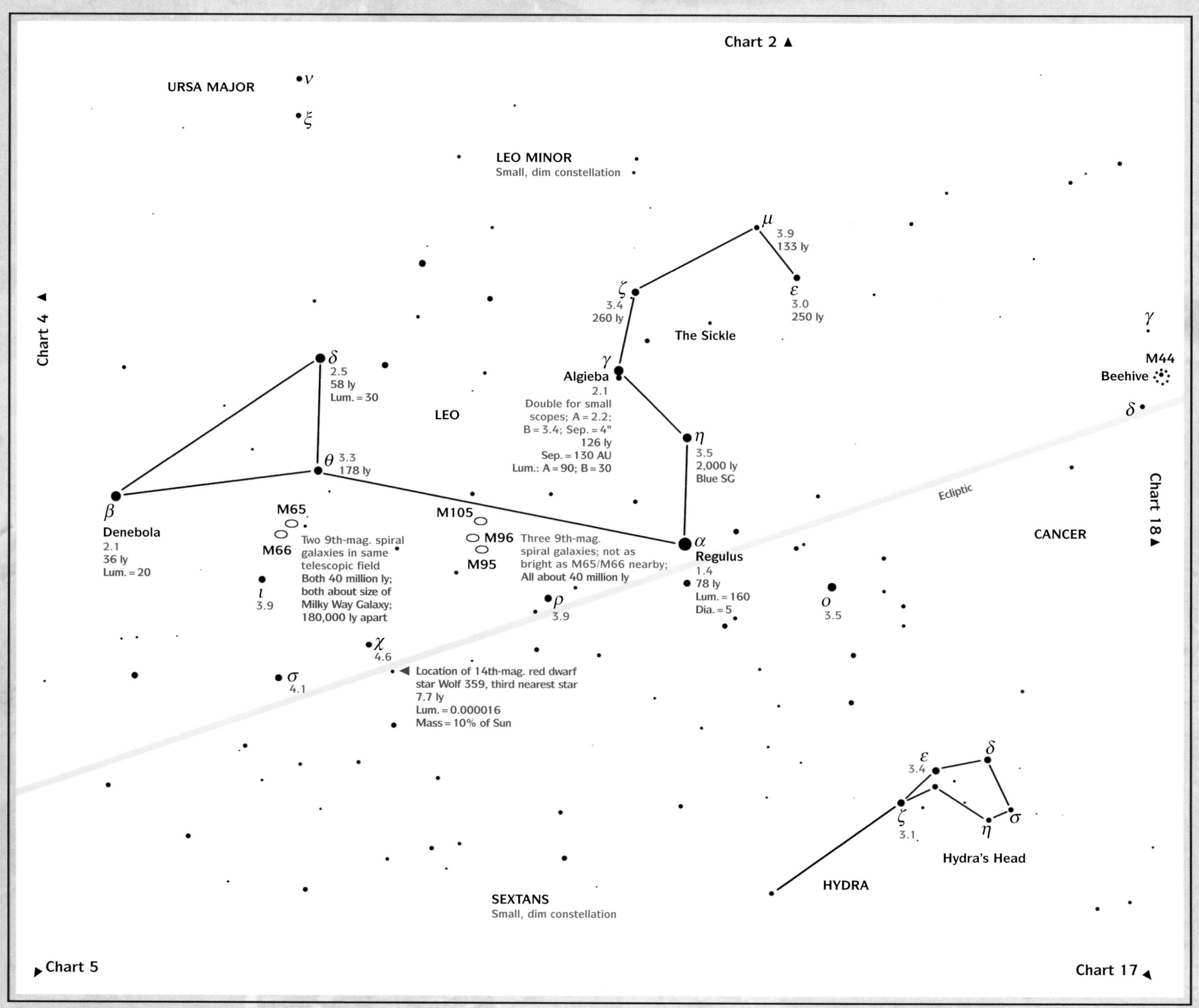

Chart 2 ▲
URSA MAJOR
LEO MINOR
Small, dim constellation
Chart 4 ◄
3.9
133 ly
3.4
260 ly
3.0
250 ly
The Sickle
2.5
58 ly
Lum. = 30
Algieba
2.1
Double for small scopes; A = 2.2; B = 3.4; Sep. = 4"
126 ly
Sep. = 130 AU
Lum.: A = 90; B = 30
LEO
M44
Beehive
3.3
178 ly
3.5
2,000 ly
Blue SG
Ecliptic
Chart 18 ►
Denebola
2.1
36 ly
Lum. = 20
M65
M66
Two 9th-mag. spiral galaxies in same telescopic field
Both 40 million ly; both about size of Milky Way Galaxy; 180,000 ly apart
M105
M96
M95
Three 9th-mag. spiral galaxies; not as bright as M65/M66 nearby; All about 40 million ly
Regulus
1.4
78 ly
Lum. = 160
Dia. = 5
CANCER
3.9
3.9
3.5
4.6
4.1
Location of 14th-mag. red dwarf star Wolf 359, third nearest star
7.7 ly
Lum. = 0.000016
Mass = 10% of Sun
3.4
3.1
Hydra's Head
HYDRA
SEXTANS
Small, dim constellation
Chart 5
Chart 17

CHART 3

LEO CANCER HYDRA

Prominently placed in spring and early-summer skies.

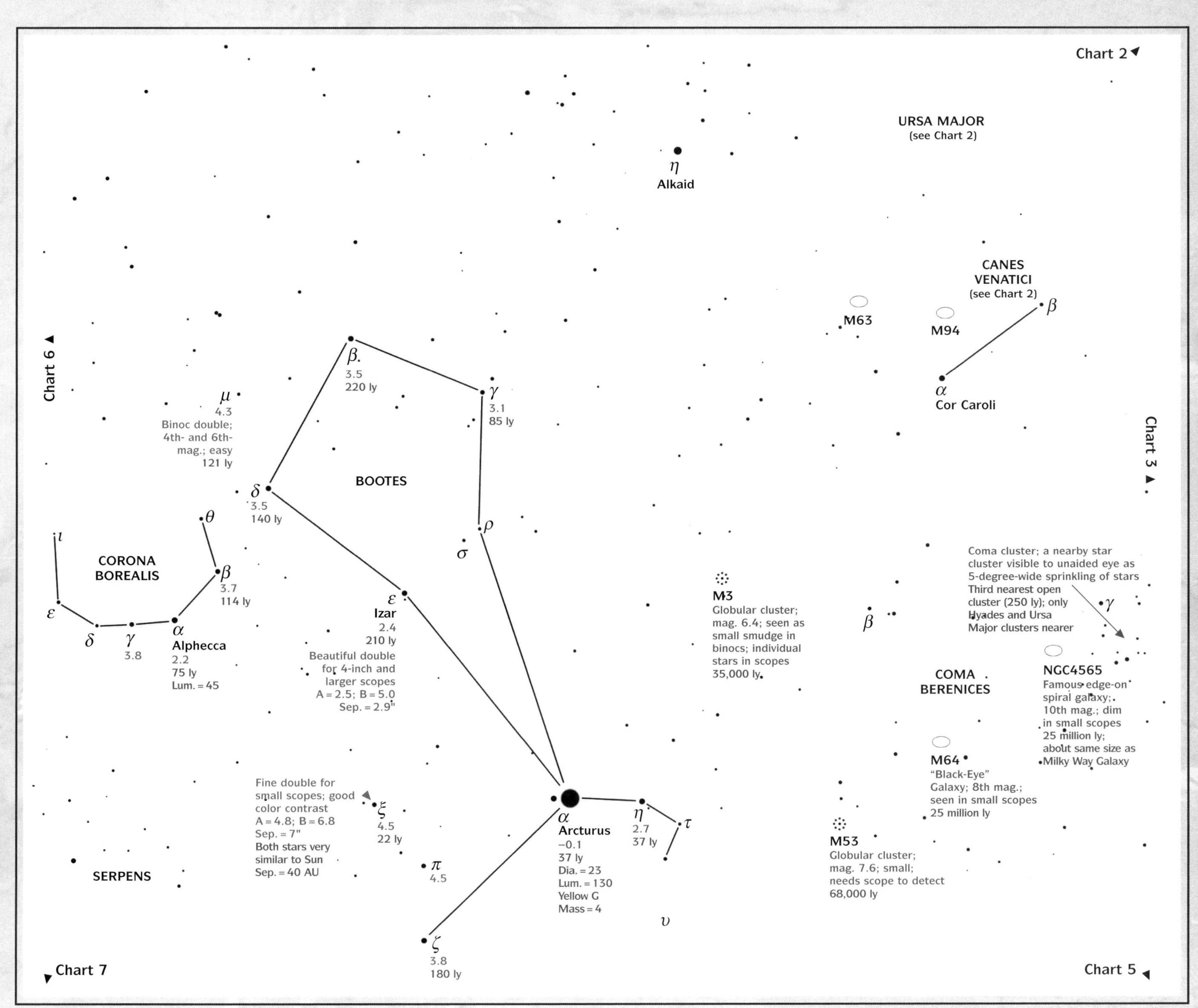

Chart 2
URSA MAJOR
(see Chart 2)
η
Alkaid
CANES
VENATICI
(see Chart 2)
β
M63
M94
α
Cor Caroli
Chart 6
Chart 3
μ
4.3
Binoc double;
4th- and 6th-
mag.; easy
121 ly
β.
3.5
220 ly
γ
3.1
85 ly
BOOTES
δ
3.5
140 ly
ρ
σ
θ
ι
CORONA
BOREALIS
β
3.7
114 ly
ε
δ
γ
3.8
α
Alphecca
2.2
75 ly
Lum. = 45
ε
Izar
2.4
210 ly
Beautiful double
for 4-inch and
larger scopes
A = 2.5; B = 5.0
Sep. = 2.9"
M3
Globular cluster;
mag. 6.4; seen as
small smudge in
binocs; individual
stars in scopes
35,000 ly
β
Coma cluster; a nearby star
cluster visible to unaided eye as
5-degree-wide sprinkling of stars
Third nearest open
cluster (250 ly); only
Hyades and Ursa
Major clusters nearer
γ
NGC4565
Famous edge-on
spiral galaxy;
10th mag.; dim
in small scopes
25 million ly;
about same size as
Milky Way Galaxy
COMA
BERENICES
M64
"Black-Eye"
Galaxy; 8th mag.;
seen in small scopes
25 million ly
Fine double for
small scopes; good
color contrast
A = 4.8; B = 6.8
Sep. = 7"
Both stars very
similar to Sun
Sep. = 40 AU
ξ
4.5
22 ly
π
4.5
α
Arcturus
−0.1
37 ly
Dia. = 23
Lum. = 130
Yellow G
Mass = 4
η
2.7
37 ly
τ
υ
M53
Globular cluster;
mag. 7.6; small;
needs scope to detect
68,000 ly
SERPENS
ζ
3.8
180 ly
Chart 7
Chart 5

CHART 4

BOOTES CORONA BOREALIS COMA BERENICES

Region high overhead in spring and early summer.

Chart 5

Virgo Corvus Hydra

Well placed in spring skies; bottom of chart close to southern horizon.

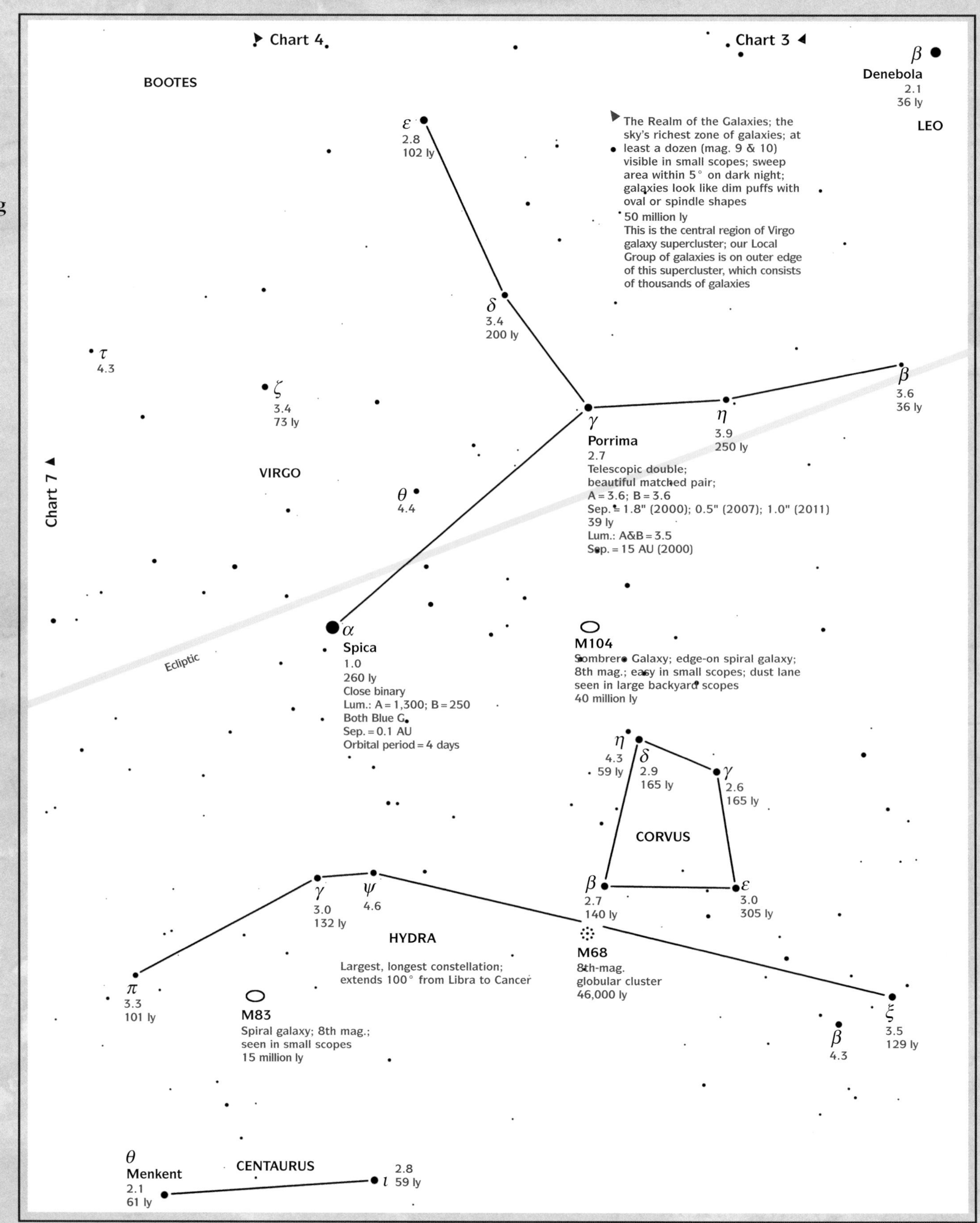
Chart 4
Chart 3
β
Denebola
2.1
36 ly
BOOTES
LEO
ε
2.8
102 ly
The Realm of the Galaxies; the sky's richest zone of galaxies; at least a dozen (mag. 9 & 10) visible in small scopes; sweep area within 5° on dark night; galaxies look like dim puffs with oval or spindle shapes
50 million ly
This is the central region of Virgo galaxy supercluster; our Local Group of galaxies is on outer edge of this supercluster, which consists of thousands of galaxies
δ
3.4
200 ly
τ
4.3
ζ
3.4
73 ly
β
3.6
36 ly
γ
Porrima
2.7
Telescopic double; beautiful matched pair;
A = 3.6; B = 3.6
Sep. = 1.8" (2000); 0.5" (2007); 1.0" (2011)
39 ly
Lum.: A&B = 3.5
Sep. = 15 AU (2000)
η
3.9
250 ly
Chart 7
VIRGO
θ
4.4
Ecliptic
α
Spica
1.0
260 ly
Close binary
Lum.: A = 1,300; B = 250
Both Blue G
Sep. = 0.1 AU
Orbital period = 4 days
M104
Sombrero Galaxy; edge-on spiral galaxy; 8th mag.; easy in small scopes; dust lane seen in large backyard scopes
40 million ly
η
4.3
59 ly
δ
2.9
165 ly
γ
2.6
165 ly
CORVUS
β
2.7
140 ly
ε
3.0
305 ly
γ
3.0
132 ly
ψ
4.6
HYDRA
Largest, longest constellation; extends 100° from Libra to Cancer
M68
8th-mag. globular cluster
46,000 ly
π
3.3
101 ly
M83
Spiral galaxy; 8th mag.; seen in small scopes
15 million ly
ξ
3.5
129 ly
β
4.3
θ
Menkent
2.1
61 ly
CENTAURUS
ι
2.8
59 ly

Chart 6

Hercules Ophiuchus Draco (head only)

Near overhead in late spring and throughout summer.

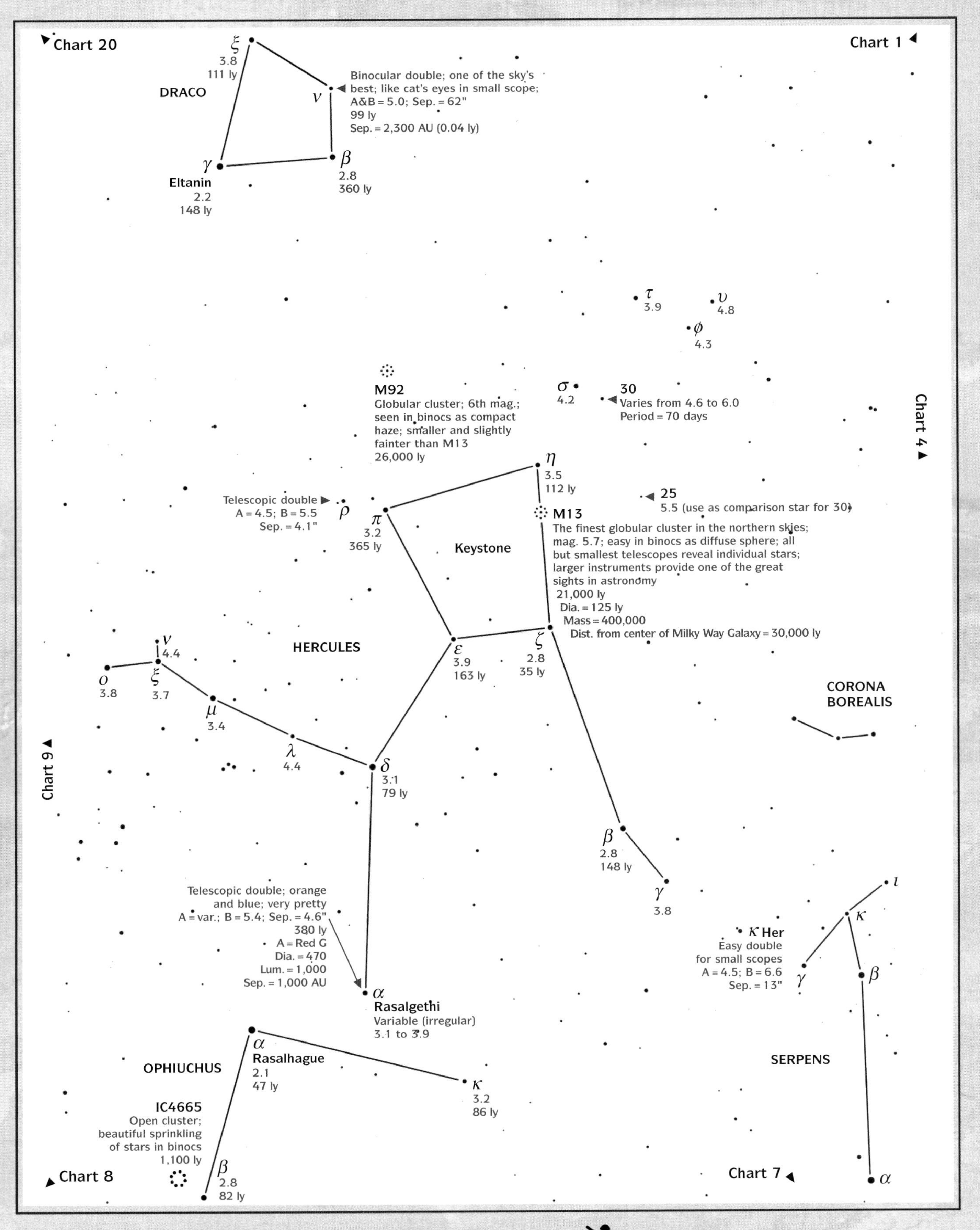

Chart 20
Chart 1
DRACO
ξ 3.8 111 ly
ν
Binocular double; one of the sky's best; like cat's eyes in small scope; A&B = 5.0; Sep. = 62" 99 ly Sep. = 2,300 AU (0.04 ly)
γ Eltanin 2.2 148 ly
β 2.8 360 ly
τ 3.9
υ 4.8
φ 4.3
M92
Globular cluster; 6th mag.; seen in binocs as compact haze; smaller and slightly fainter than M13
26,000 ly
σ 4.2
30
Varies from 4.6 to 6.0
Period = 70 days
Chart 4
η 3.5 112 ly
25
5.5 (use as comparison star for 30)
Telescopic double
A = 4.5; B = 5.5
Sep. = 4.1"
ρ
π 3.2 365 ly
Keystone
M13
The finest globular cluster in the northern skies; mag. 5.7; easy in binocs as diffuse sphere; all but smallest telescopes reveal individual stars; larger instruments provide one of the great sights in astronomy
21,000 ly
Dia. = 125 ly
Mass = 400,000
Dist. from center of Milky Way Galaxy = 30,000 ly
HERCULES
ν 4.4
ο 3.8
ξ 3.7
μ 3.4
λ 4.4
δ 3.1 79 ly
ε 3.9 163 ly
ζ 2.8 35 ly
CORONA BOREALIS
Chart 9
β 2.8 148 ly
γ 3.8
ι
κ
Telescopic double; orange and blue; very pretty
A = var.; B = 5.4; Sep. = 4.6"
380 ly
A = Red G
Dia. = 470
Lum. = 1,000
Sep. = 1,000 AU
κ Her
Easy double for small scopes
A = 4.5; B = 6.6
Sep. = 13"
γ
β
α Rasalgethi
Variable (irregular)
3.1 to 3.9
α Rasalhague 2.1 47 ly
κ 3.2 86 ly
OPHIUCHUS
SERPENS
IC4665
Open cluster; beautiful sprinkling of stars in binocs
1,100 ly
β 2.8 82 ly
Chart 8
Chart 7
α

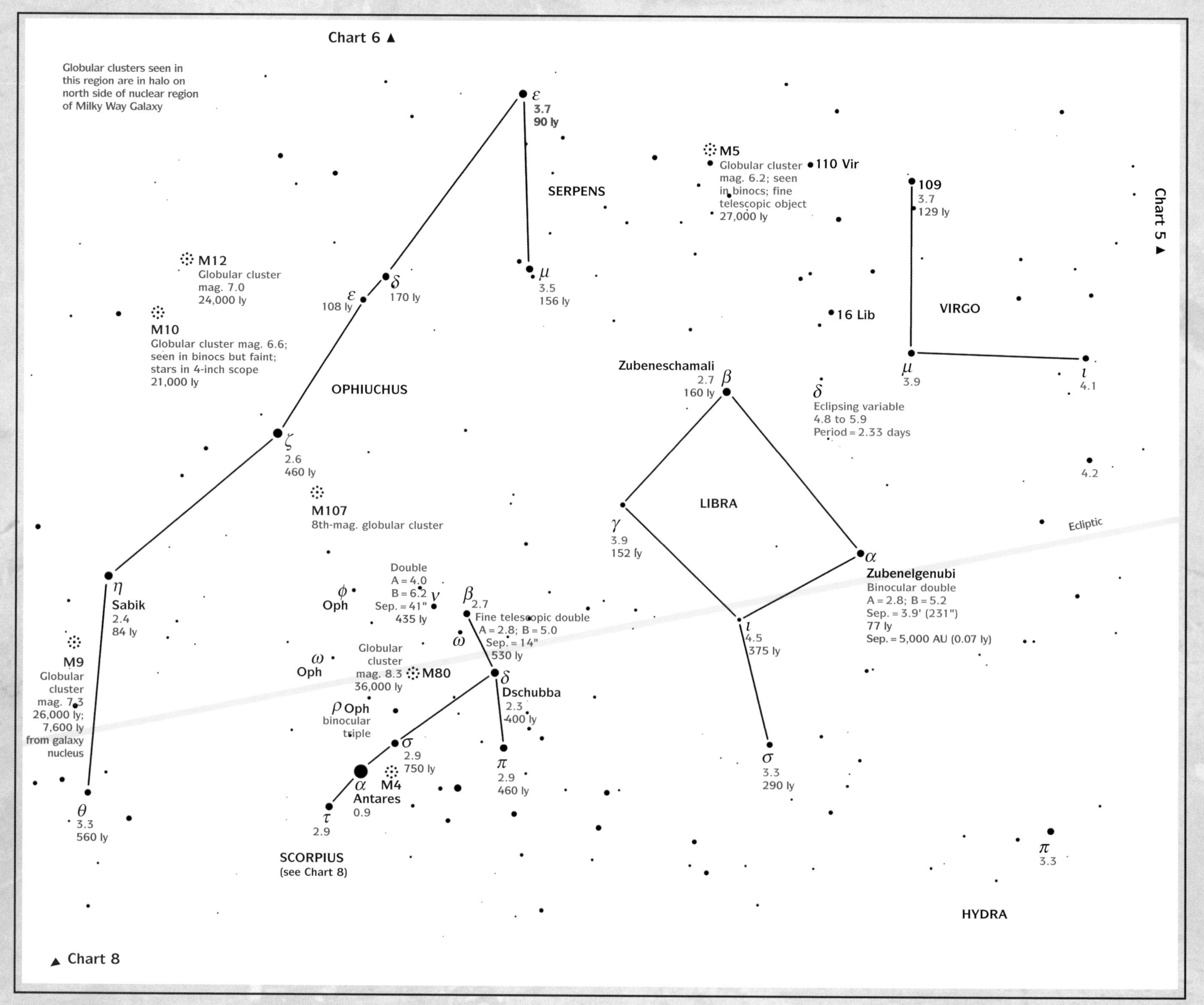
Chart 6
Globular clusters seen in this region are in halo on north side of nuclear region of Milky Way Galaxy
ε 3.7 90 ly
SERPENS
M5 Globular cluster mag. 6.2; seen in binocs; fine telescopic object 27,000 ly
110 Vir
109 3.7 129 ly
Chart 5
M12 Globular cluster mag. 7.0 24,000 ly
M10 Globular cluster mag. 6.6; seen in binocs but faint; stars in 4-inch scope 21,000 ly
ε 108 ly
δ 170 ly
μ 3.5 156 ly
16 Lib
VIRGO
OPHIUCHUS
Zubeneschamali 2.7 160 ly
μ 3.9
ι 4.1
δ Eclipsing variable 4.8 to 5.9 Period = 2.33 days
ζ 2.6 460 ly
4.2
M107 8th-mag. globular cluster
LIBRA
γ 3.9 152 ly
Ecliptic
Zubenelgenubi Binocular double A = 2.8; B = 5.2 Sep. = 3.9' (231") 77 ly Sep. = 5,000 AU (0.07 ly)
η Sabik 2.4 84 ly
Double A = 4.0 B = 6.2 Sep. = 41" 435 ly
φ Oph
β 2.7 Fine telescopic double A = 2.8; B = 5.0 Sep. = 14" 530 ly
ι 4.5 375 ly
M9 Globular cluster mag. 7.3 26,000 ly; 7,600 ly from galaxy nucleus
ω Oph
Globular cluster mag. 8.3 36,000 ly M80
δ Dschubba 2.3 400 ly
ρ Oph binocular triple
σ 2.9 750 ly
π 2.9 460 ly
σ 3.3 290 ly
α Antares 0.9
M4
τ 2.9
θ 3.3 560 ly
π 3.3
SCORPIUS (see Chart 8)
HYDRA
Chart 8

CHART 7

OPHIUCHUS LIBRA SCORPIUS (NORTHERN PART)

High in southern sky throughout summer months.

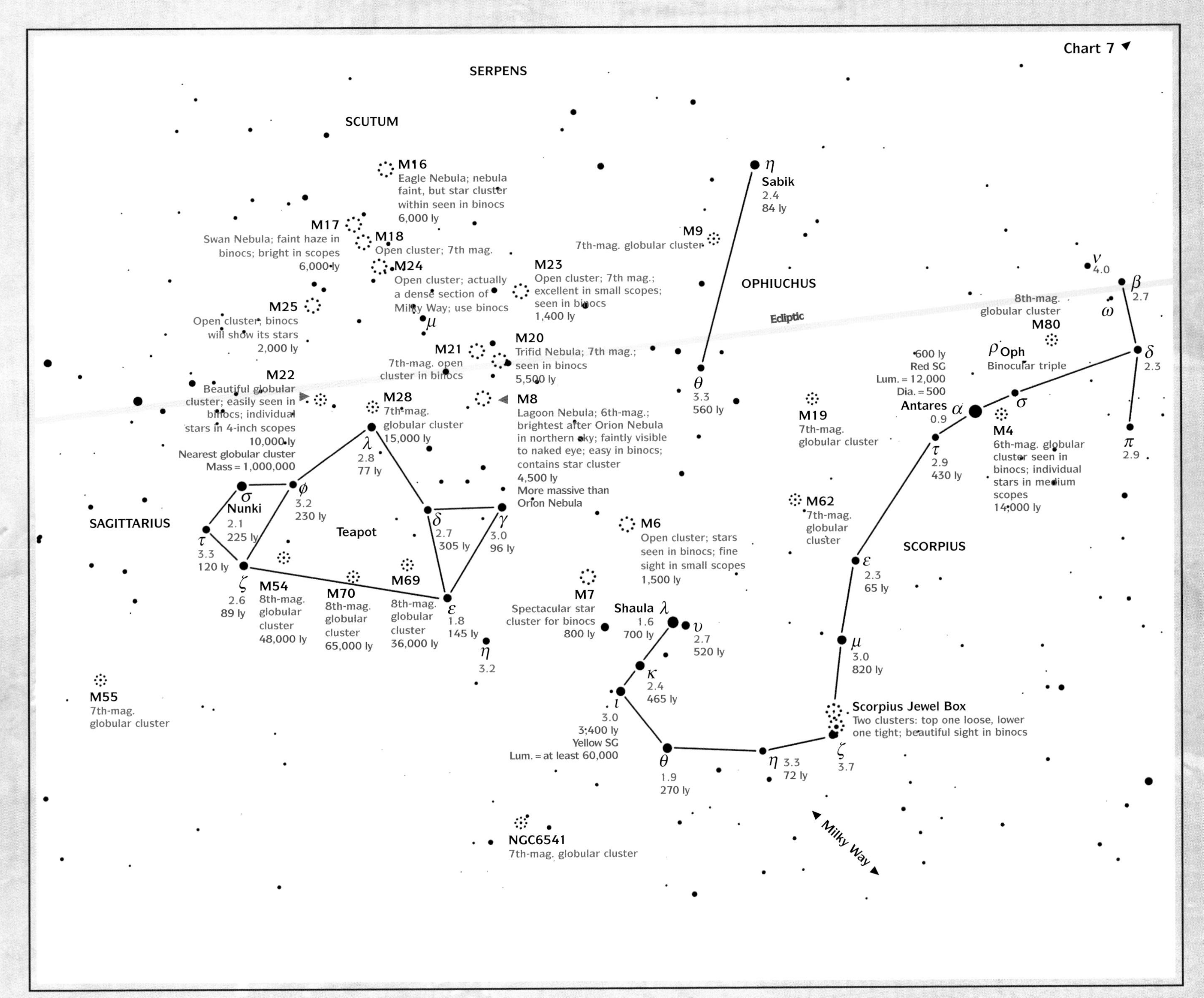

Chart 7
SERPENS
SCUTUM
M16
Eagle Nebula; nebula faint, but star cluster within seen in binocs
6,000 ly
M17
Swan Nebula; faint haze in binocs; bright in scopes
6,000 ly
M18
Open cluster; 7th mag.
M24
Open cluster; actually a dense section of Milky Way; use binocs
μ
M23
Open cluster; 7th mag.; excellent in small scopes; seen in binocs
1,400 ly
M25
Open cluster; binocs will show its stars
2,000 ly
M22
Beautiful globular cluster; easily seen in binocs; individual stars in 4-inch scopes
10,000 ly
Nearest globular cluster
Mass = 1,000,000
M21
7th-mag. open cluster in binocs
M20
Trifid Nebula; 7th mag.; seen in binocs
5,500 ly
M28
7th-mag. globular cluster
15,000 ly
M8
Lagoon Nebula; 6th-mag.; brightest after Orion Nebula in northern sky; faintly visible to naked eye; easy in binocs; contains star cluster
4,500 ly
More massive than Orion Nebula
M9
7th-mag. globular cluster
η
Sabik
2.4
84 ly
OPHIUCHUS
Ecliptic
θ
3.3
560 ly
SAGITTARIUS
σ
Nunki
2.1
225 ly
φ
3.2
230 ly
λ
2.8
77 ly
τ
3.3
120 ly
Teapot
δ
2.7
305 ly
γ
3.0
96 ly
ζ
2.6
89 ly
M54
8th-mag. globular cluster
48,000 ly
M70
8th-mag. globular cluster
65,000 ly
M69
8th-mag. globular cluster
36,000 ly
ε
1.8
145 ly
η
3.2
M55
7th-mag. globular cluster
M6
Open cluster; stars seen in binocs; fine sight in small scopes
1,500 ly
M7
Spectacular star cluster for binocs
800 ly
Shaula λ
1.6
700 ly
υ
2.7
520 ly
κ
2.4
465 ly
ι
3.0
3,400 ly
Yellow SG
Lum. = at least 60,000
θ
1.9
270 ly
η
3.3
72 ly
ζ
3.7
Scorpius Jewel Box
Two clusters: top one loose, lower one tight; beautiful sight in binocs
NGC6541
7th-mag. globular cluster
Milky Way
M19
7th-mag. globular cluster
M62
7th-mag. globular cluster
SCORPIUS
ε
2.3
65 ly
μ
3.0
820 ly
τ
2.9
430 ly
600 ly
Red SG
Lum. = 12,000
Dia. = 500
Antares α
0.9
σ
ρ Oph
Binocular triple
M4
6th-mag. globular cluster seen in binocs; individual stars in medium scopes
14,000 ly
8th-mag. globular cluster
M80
ν
4.0
β
2.7
ω
δ
2.3
π
2.9

CHART 8

SCORPIUS SAGITTARIUS SCUTUM

Richest zone of Milky Way carves through left side of this region; seen near southern horizon mid- to late summer.

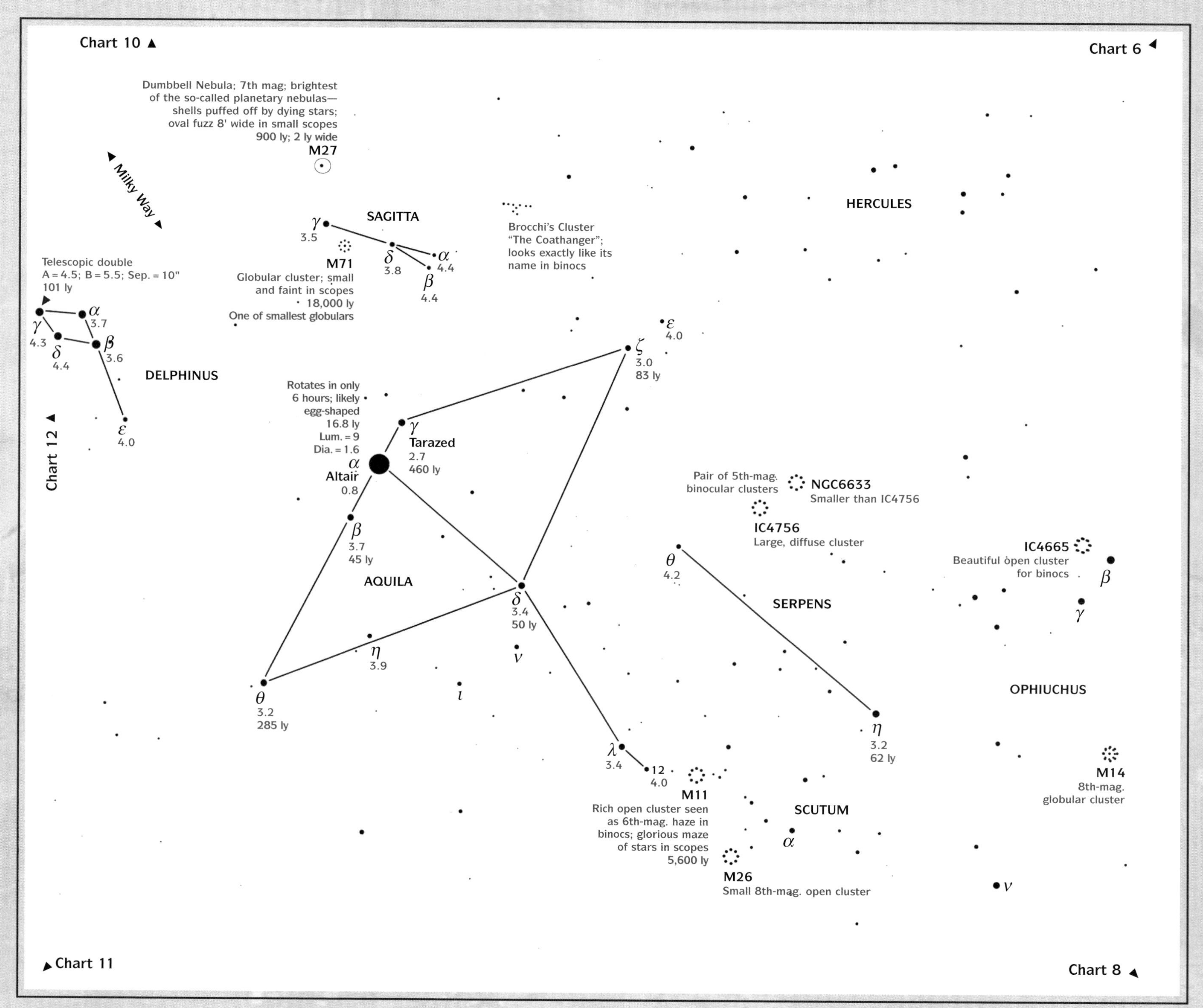

CHART 9

AQUILA DELPHINUS SAGITTA

Bottom part of Summer Triangle; seen throughout summer and early autumn.

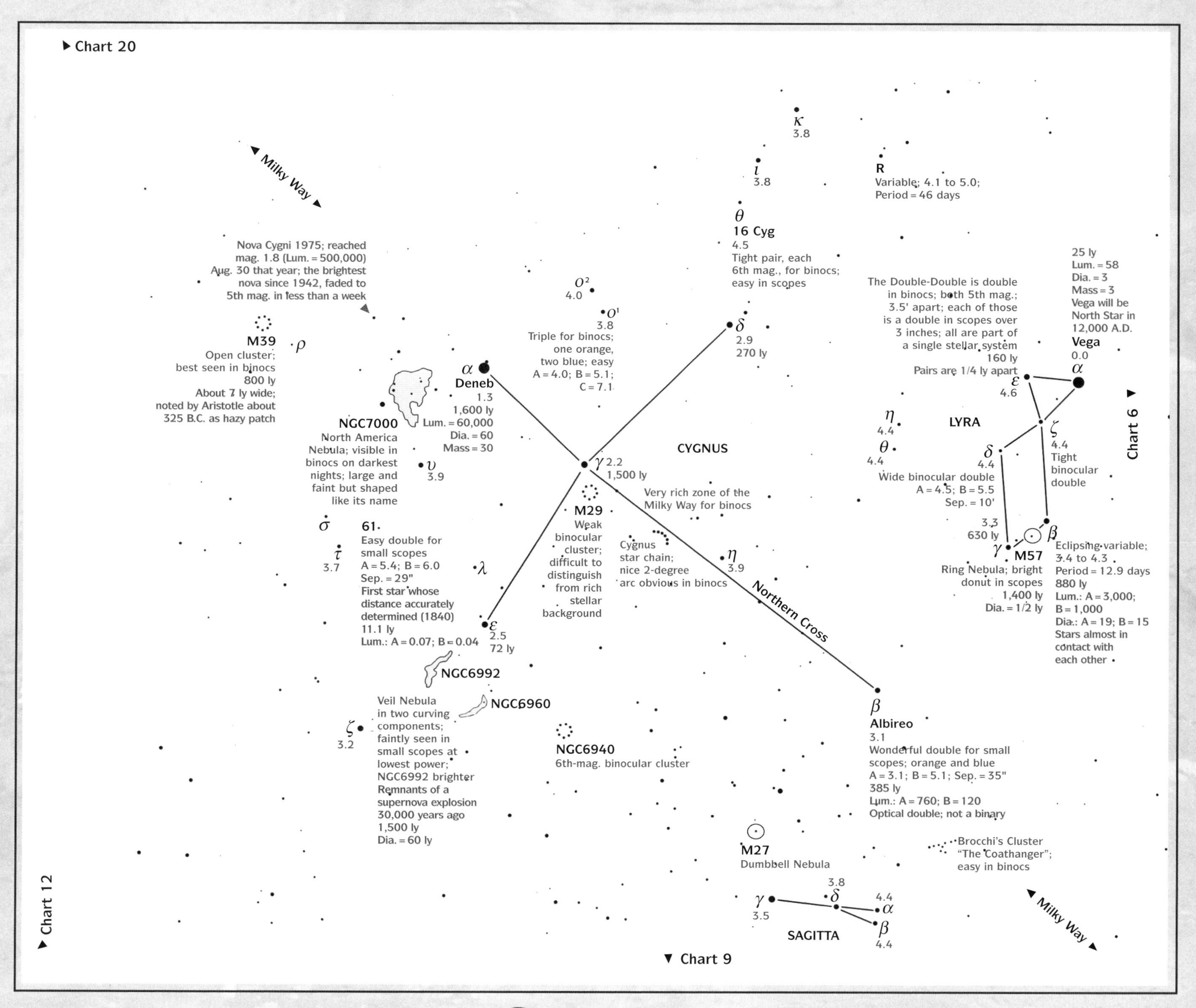

Chart 20
Milky Way
Nova Cygni 1975; reached mag. 1.8 (Lum. = 500,000) Aug. 30 that year; the brightest nova since 1942, faded to 5th mag. in less than a week
M39
Open cluster; best seen in binocs
800 ly
About 7 ly wide; noted by Aristotle about 325 B.C. as hazy patch
ρ
NGC7000
North America Nebula; visible in binocs on darkest nights; large and faint but shaped like its name
υ
3.9
α
Deneb
1.3
1,600 ly
Lum. = 60,000
Dia. = 60
Mass = 30
ο²
4.0
ο¹
3.8
Triple for binocs; one orange, two blue; easy
A = 4.0; B = 5.1; C = 7.1
κ
3.8
ι
3.8
θ
16 Cyg
4.5
Tight pair, each 6th mag., for binocs; easy in scopes
δ
2.9
270 ly
R
Variable; 4.1 to 5.0; Period = 46 days
CYGNUS
γ 2.2
1,500 ly
Very rich zone of the Milky Way for binocs
M29
Weak binocular cluster; difficult to distinguish from rich stellar background
Cygnus star chain; nice 2-degree arc obvious in binocs
η
3.9
Northern Cross
σ
τ
3.7
61
Easy double for small scopes
A = 5.4; B = 6.0
Sep. = 29"
First star whose distance accurately determined (1840)
11.1 ly
Lum.: A = 0.07; B = 0.04
λ
ε
2.5
72 ly
NGC6992
NGC6960
ζ
3.2
Veil Nebula in two curving components; faintly seen in small scopes at lowest power; NGC6992 brighter
Remnants of a supernova explosion 30,000 years ago
1,500 ly
Dia. = 60 ly
NGC6940
6th-mag. binocular cluster
β
Albireo
3.1
Wonderful double for small scopes; orange and blue
A = 3.1; B = 5.1; Sep. = 35"
385 ly
Lum.: A = 760; B = 120
Optical double; not a binary
M27
Dumbbell Nebula
Brocchi's Cluster "The Coathanger"; easy in binocs
γ
3.5
δ
3.8
α
4.4
β
4.4
SAGITTA
Milky Way
Chart 9
Chart 12
The Double-Double is double in binocs; both 5th mag.; 3.5' apart; each of those is a double in scopes over 3 inches; all are part of a single stellar system
160 ly
Pairs are 1/4 ly apart
ε
4.6
25 ly
Lum. = 58
Dia. = 3
Mass = 3
Vega will be North Star in 12,000 A.D.
Vega
0.0
α
Chart 6
η
4.4
θ
4.4
LYRA
δ
4.4
Wide binocular double
A = 4.5; B = 5.5
Sep. = 10'
ζ
4.4
Tight binocular double
3.3
630 ly
γ
M57
β
Ring Nebula; bright donut in scopes
1,400 ly
Dia. = 1/2 ly
Eclipsing variable; 3.4 to 4.3
Period = 12.9 days
880 ly
Lum.: A = 3,000; B = 1,000
Dia.: A = 19; B = 15
Stars almost in contact with each other

Chart 10

Cygnus Lyra

Main section of Summer Triangle;
in northeast in spring, overhead in summer, in northwest in autumn.

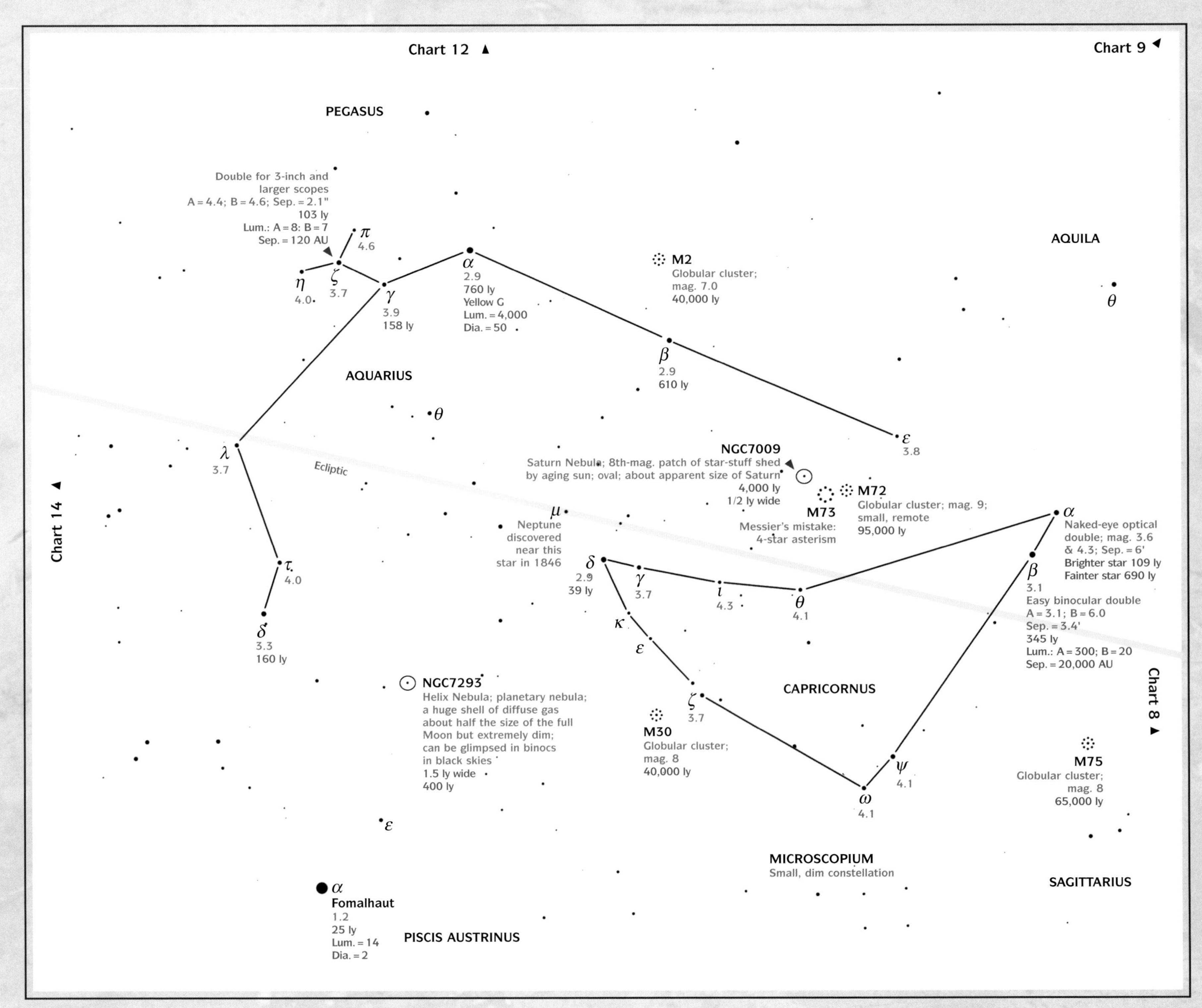
Chart 12
Chart 9
PEGASUS
Double for 3-inch and larger scopes A = 4.4; B = 4.6; Sep. = 2.1" 103 ly Lum.: A = 8: B = 7 Sep. = 120 AU
π 4.6
η 4.0
ζ 3.7
γ 3.9 158 ly
α 2.9 760 ly Yellow G Lum. = 4,000 Dia. = 50
M2
Globular cluster; mag. 7.0 40,000 ly
AQUILA
θ
β 2.9 610 ly
AQUARIUS
θ
ε 3.8
λ 3.7
Ecliptic
NGC7009
Saturn Nebula; 8th-mag. patch of star-stuff shed by aging sun; oval; about apparent size of Saturn 4,000 ly 1/2 ly wide
M72
Globular cluster; mag. 9; small, remote 95,000 ly
M73
Messier's mistake: 4-star asterism
α
Naked-eye optical double; mag. 3.6 & 4.3; Sep. = 6' Brighter star 109 ly Fainter star 690 ly
Chart 14
μ
Neptune discovered near this star in 1846
τ 4.0
δ 2.9 39 ly
γ 3.7
ι 4.3
θ 4.1
β 3.1
Easy binocular double A = 3.1; B = 6.0 Sep. = 3.4' 345 ly Lum.: A = 300; B = 20 Sep. = 20,000 AU
δ 3.3 160 ly
κ
ε
NGC7293
Helix Nebula; planetary nebula; a huge shell of diffuse gas about half the size of the full Moon but extremely dim; can be glimpsed in binocs in black skies 1.5 ly wide 400 ly
ζ 3.7
CAPRICORNUS
Chart 8
M30
Globular cluster; mag. 8 40,000 ly
ψ 4.1
ω 4.1
M75
Globular cluster; mag. 8 65,000 ly
ε
MICROSCOPIUM
Small, dim constellation
SAGITTARIUS
α Fomalhaut 1.2 25 ly Lum. = 14 Dia. = 2
PISCIS AUSTRINUS

CHART 11

CAPRICORNUS AQUARIUS

Two dim zodiac constellations seen in south in autumn.

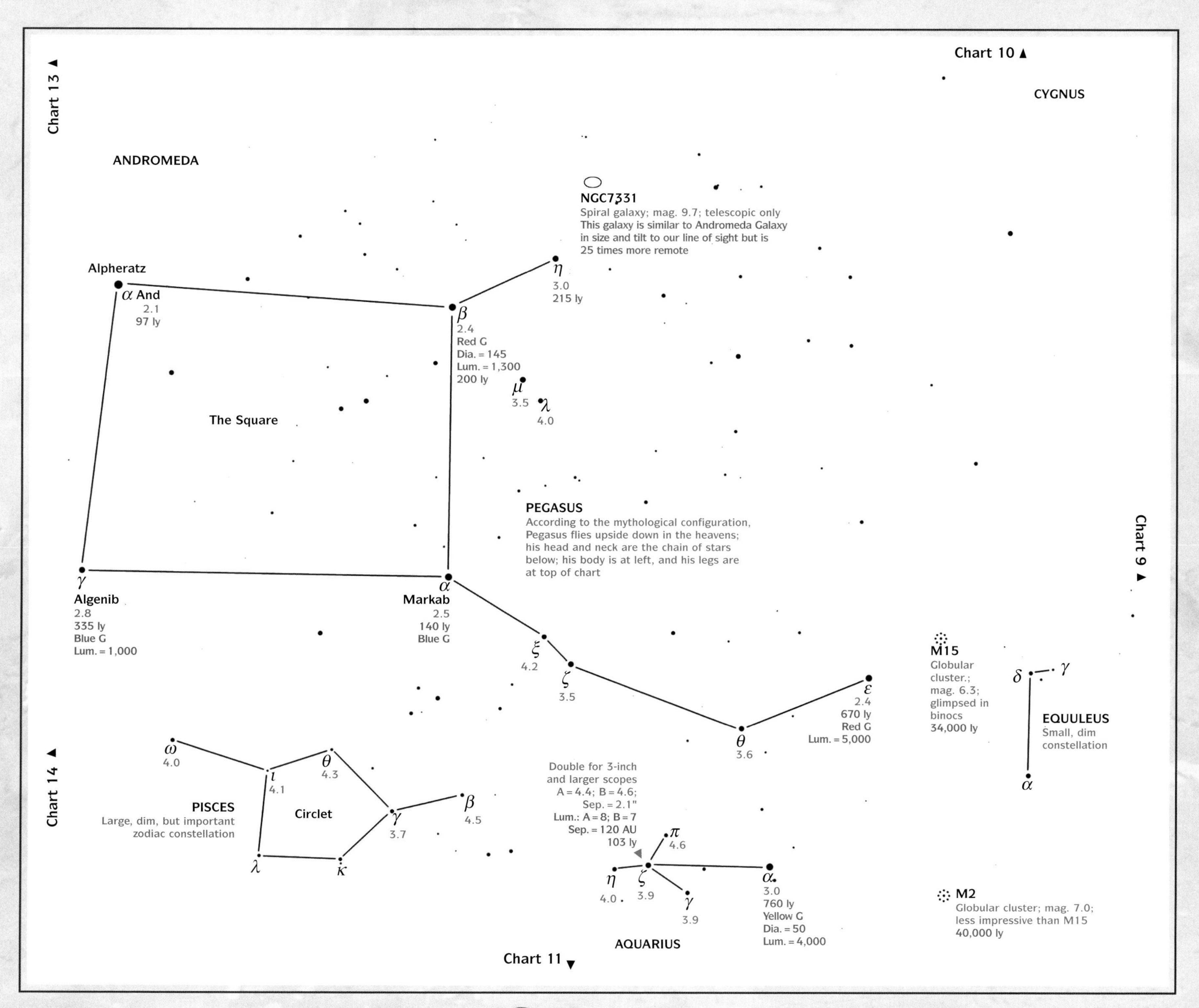

Chart 12

Pegasus Pisces (part) Aquarius (part)

A major guidepost region to autumn skies;
near overhead October to December.

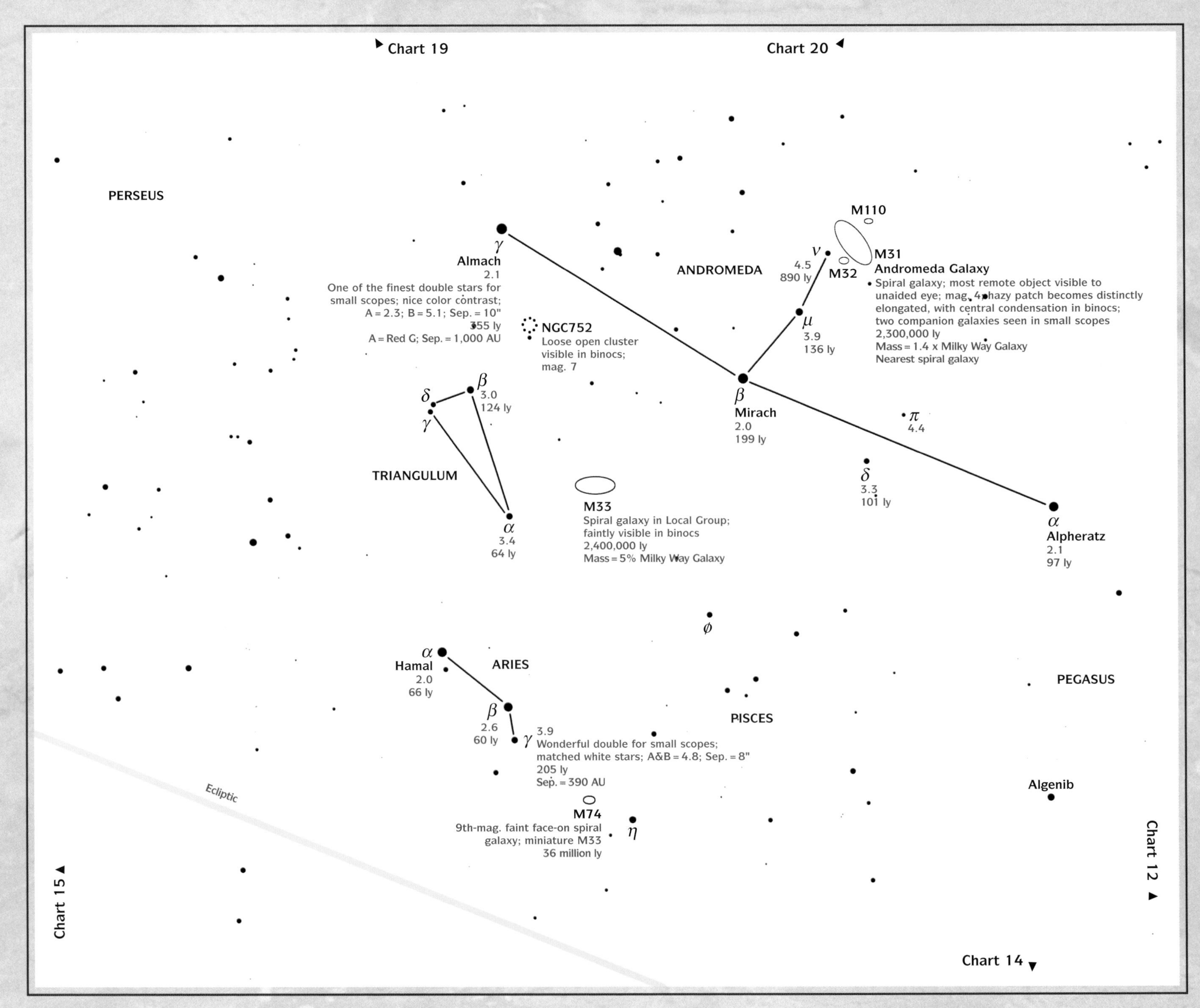

Chart 13

Andromeda Aries Triangulum

In northeast sky in early autumn, overhead late autumn, in northwest early winter.

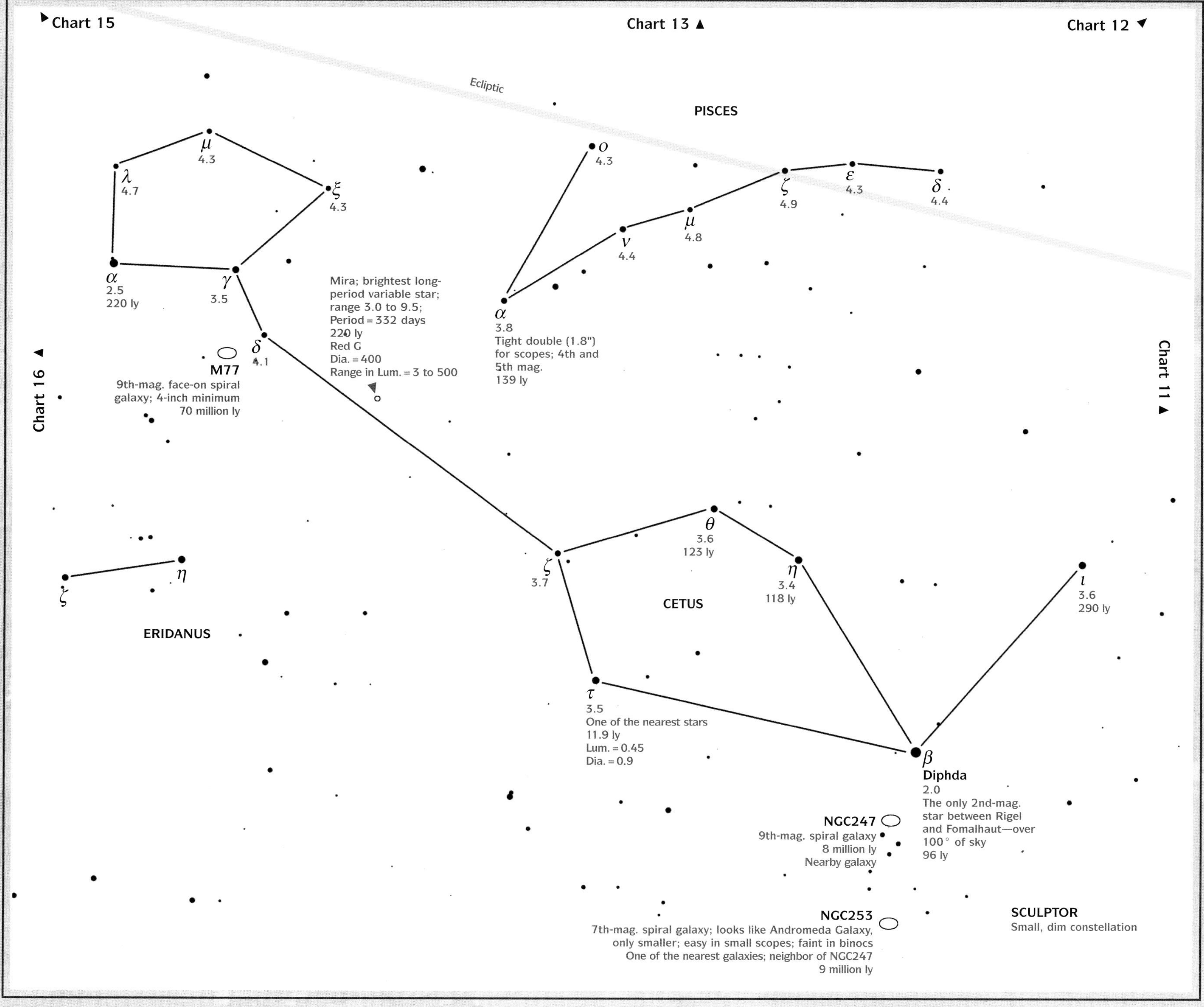

Chart 15
Chart 13
Chart 12
Ecliptic
PISCES
μ
4.3
λ
4.7
ξ
4.3
α
2.5
220 ly
γ
3.5
Mira; brightest long-period variable star; range 3.0 to 9.5; Period = 332 days
220 ly
Red G
Dia. = 400
Range in Lum. = 3 to 500
δ
4.1
M77
9th-mag. face-on spiral galaxy; 4-inch minimum
70 million ly
Chart 16
o
4.3
ζ
4.9
ε
4.3
δ
4.4
μ
4.8
ν
4.4
α
3.8
Tight double (1.8") for scopes; 4th and 5th mag.
139 ly
Chart 11
θ
3.6
123 ly
ζ
3.7
η
3.4
118 ly
ι
3.6
290 ly
CETUS
ζ
η
ERIDANUS
τ
3.5
One of the nearest stars
11.9 ly
Lum. = 0.45
Dia. = 0.9
β
Diphda
2.0
The only 2nd-mag. star between Rigel and Fomalhaut—over 100° of sky
96 ly
NGC247
9th-mag. spiral galaxy
8 million ly
Nearby galaxy
NGC253
7th-mag. spiral galaxy; looks like Andromeda Galaxy, only smaller; easy in small scopes; faint in binocs
One of the nearest galaxies; neighbor of NGC247
9 million ly
SCULPTOR
Small, dim constellation

Chart 14

Cetus Pisces

Dim constellations of the autumn southern sky.

CHART 15

TAURUS
AURIGA
ORION

Three of the six major constellations of the winter sky.

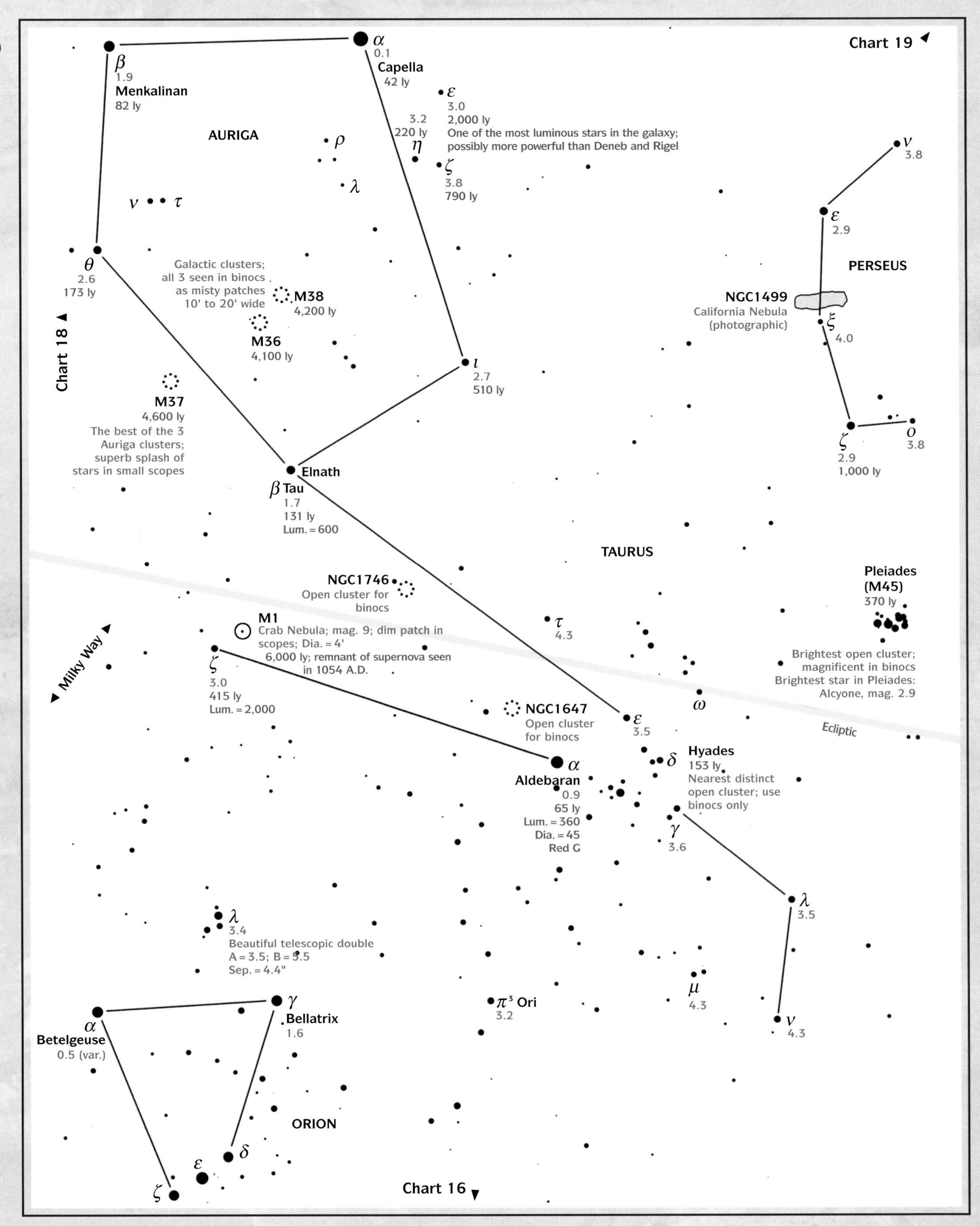

Chart 19
α
0.1
Capella
42 ly
β
1.9
Menkalinan
82 ly
ε
3.0
2,000 ly
One of the most luminous stars in the galaxy; possibly more powerful than Deneb and Rigel
3.2
220 ly
η
AURIGA
ρ
ζ
3.8
790 ly
λ
ν
τ
ν
3.8
ε
2.9
PERSEUS
θ
2.6
173 ly
Galactic clusters; all 3 seen in binocs as misty patches 10' to 20' wide
M38
4,200 ly
M36
4,100 ly
NGC1499
California Nebula (photographic)
ξ
4.0
Chart 18
M37
4,600 ly
The best of the 3 Auriga clusters; superb splash of stars in small scopes
ι
2.7
510 ly
ζ
2.9
1,000 ly
o
3.8
Elnath
β Tau
1.7
131 ly
Lum. = 600
TAURUS
NGC1746
Open cluster for binocs
Pleiades (M45)
370 ly
M1
Crab Nebula; mag. 9; dim patch in scopes; Dia. = 4'
6,000 ly; remnant of supernova seen in 1054 A.D.
τ
4.3
Milky Way
ζ
3.0
415 ly
Lum. = 2,000
Brightest open cluster; magnificent in binocs
Brightest star in Pleiades: Alcyone, mag. 2.9
NGC1647
Open cluster for binocs
ε
3.5
ω
Ecliptic
Hyades
153 ly
Nearest distinct open cluster; use binocs only
δ
α
Aldebaran
0.9
65 ly
Lum. = 360
Dia. = 45
Red G
γ
3.6
λ
3.5
λ
3.4
Beautiful telescopic double
A = 3.5; B = 5.5
Sep. = 4.4"
μ
4.3
γ
Bellatrix
1.6
α
Betelgeuse
0.5 (var.)
π³ Ori
3.2
ν
4.3
ORION
δ
ε
ζ
Chart 16

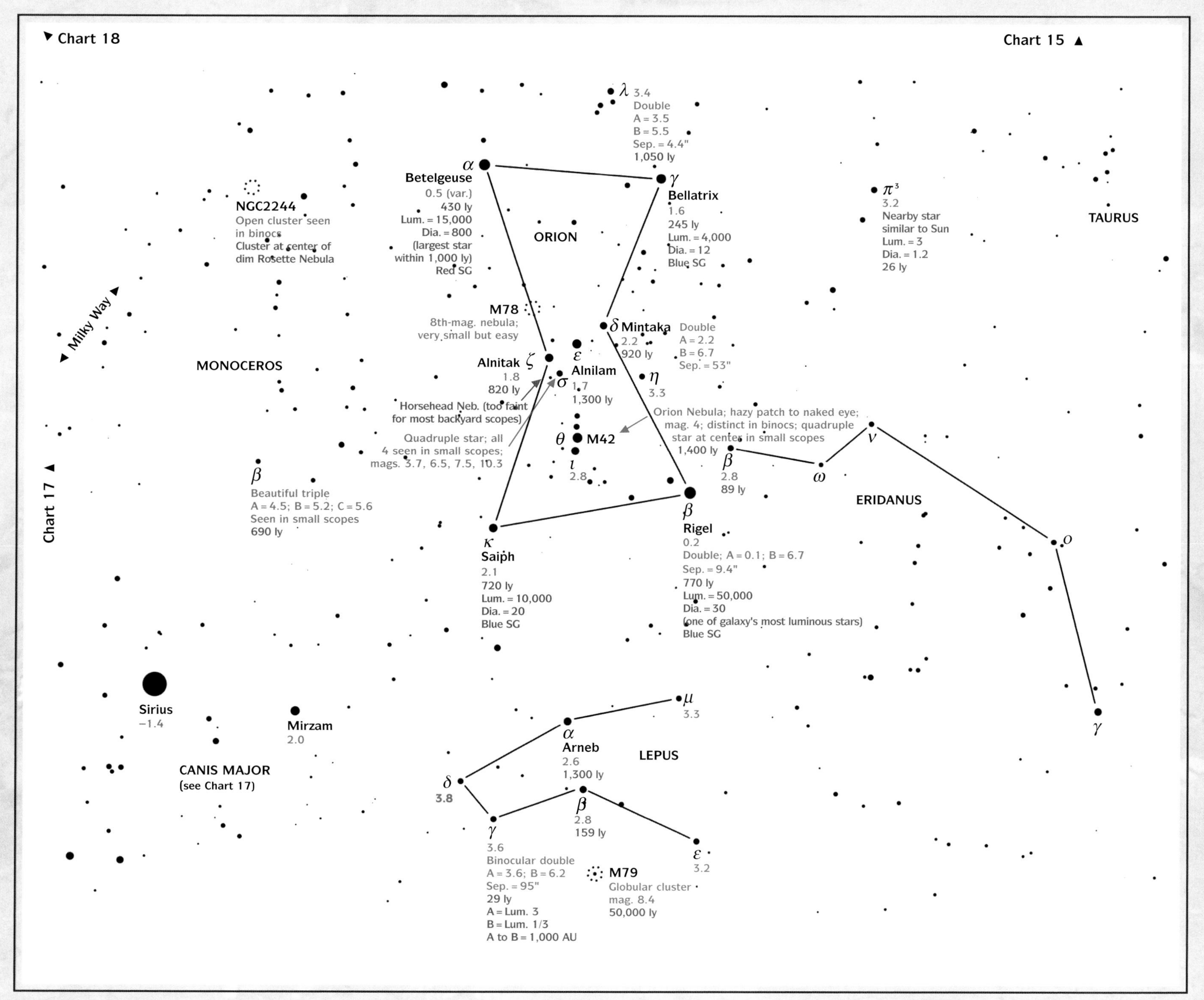

Chart 16

Orion Lepus Monoceros

Orion is the key to the winter sky,
in addition to housing a vast array of celestial sights.

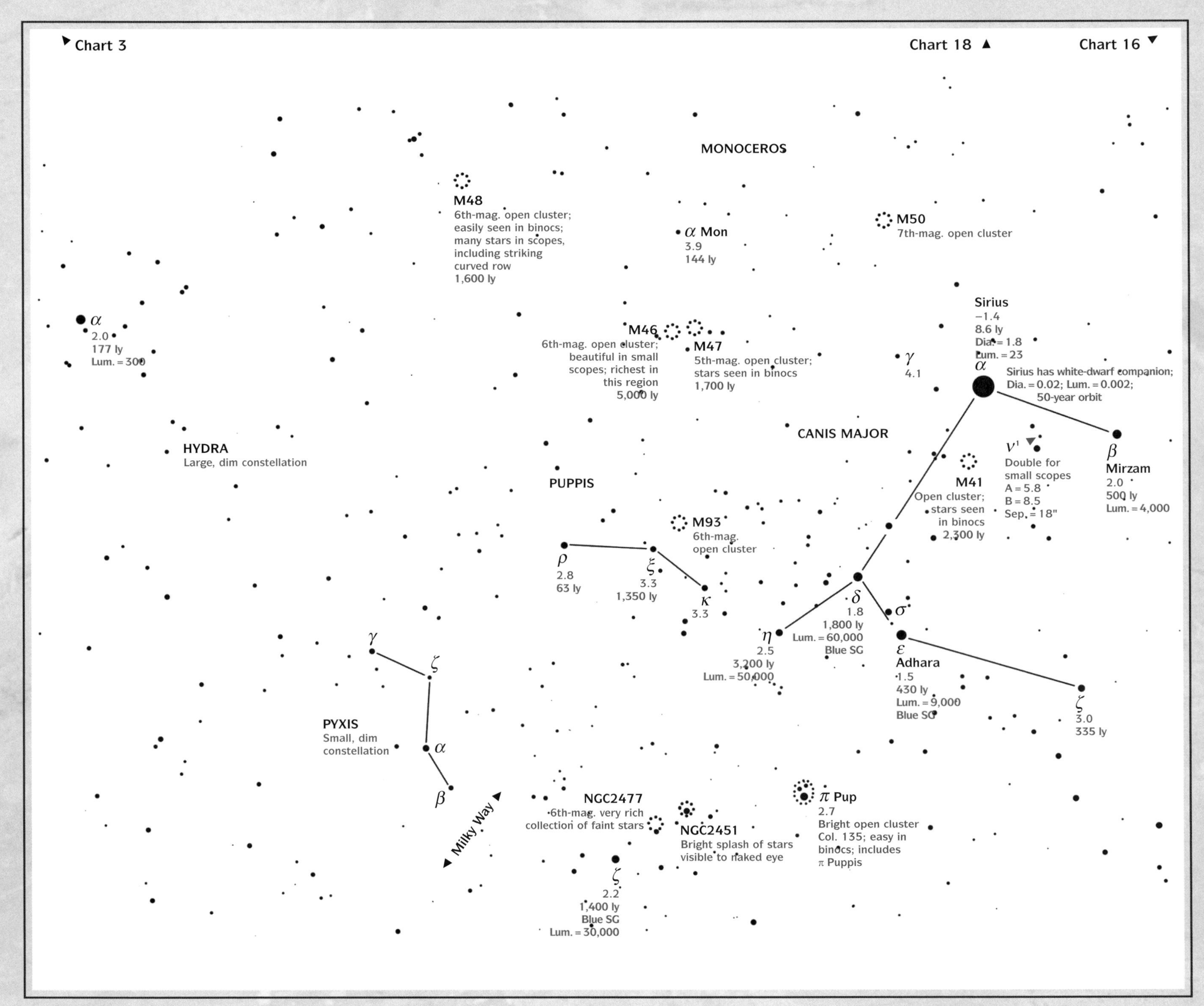

Chart 17

Canis Major Puppis Monoceros

Sirius, the night sky's brightest star, highlights this late-winter/early-spring sector of the southern sky.

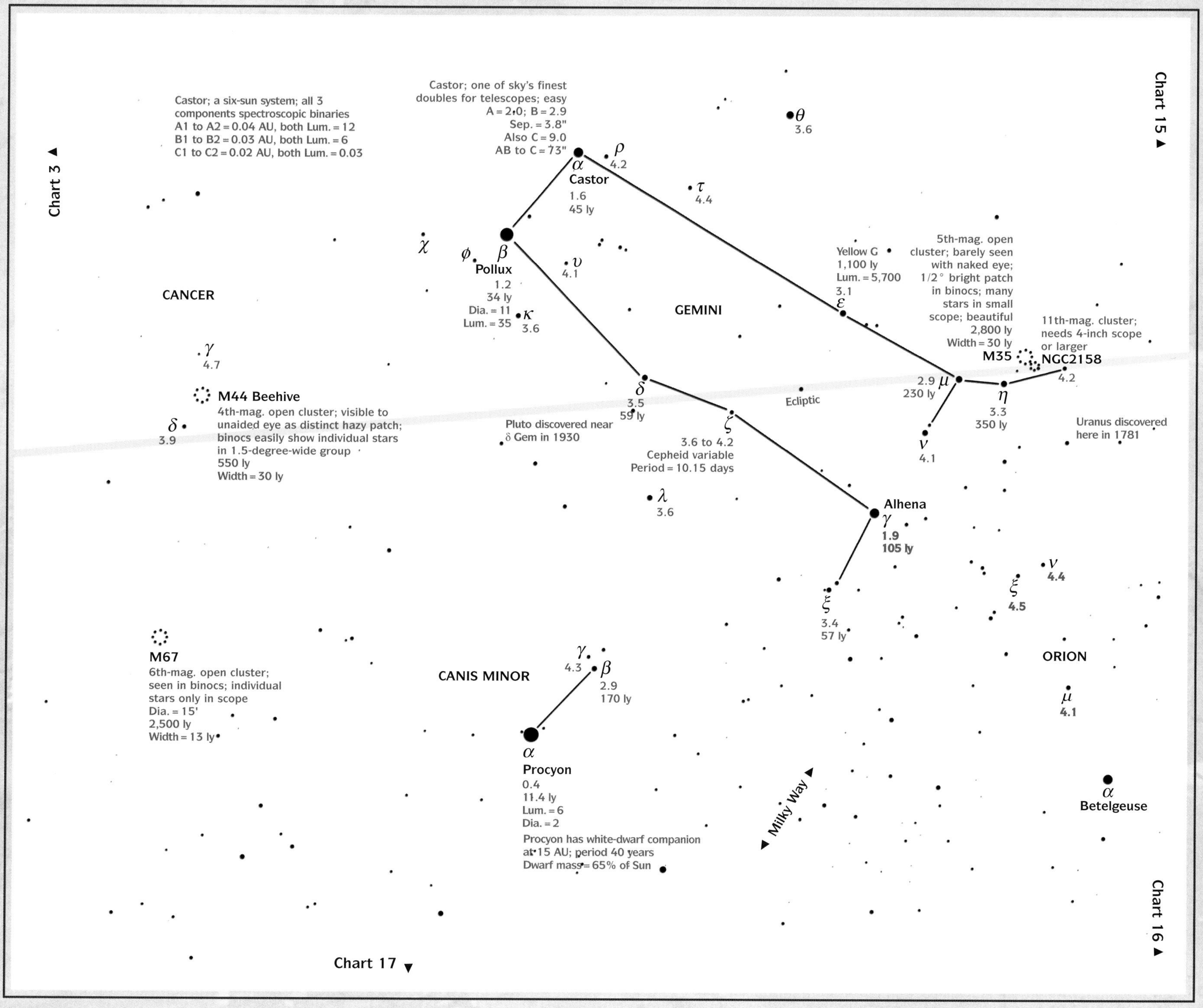

Castor; a six-sun system; all 3 components spectroscopic binaries
A1 to A2 = 0.04 AU, both Lum. = 12
B1 to B2 = 0.03 AU, both Lum. = 6
C1 to C2 = 0.02 AU, both Lum. = 0.03

Castor; one of sky's finest doubles for telescopes; easy
A = 2.0; B = 2.9
Sep. = 3.8"
Also C = 9.0
AB to C = 73"

Chart 3 ◀

Chart 15 ▶

θ 3.6

ρ 4.2

α Castor 1.6 45 ly

τ 4.4

χ

φ

β Pollux 1.2 34 ly Dia. = 11 Lum. = 35

υ 4.1

κ 3.6

CANCER

GEMINI

Yellow G
1,100 ly
Lum. = 5,700
3.1
ε

5th-mag. open cluster; barely seen with naked eye; 1/2° bright patch in binocs; many stars in small scope; beautiful
2,800 ly
Width = 30 ly

11th-mag. cluster; needs 4-inch scope or larger

M35

NGC2158

γ 4.7

M44 Beehive

4th-mag. open cluster; visible to unaided eye as distinct hazy patch; binocs easily show individual stars in 1.5-degree-wide group
550 ly
Width = 30 ly

δ 3.9

δ 3.5 59 ly

Pluto discovered near δ Gem in 1930

ζ
3.6 to 4.2
Cepheid variable
Period = 10.15 days

Ecliptic

2.9 μ 230 ly

η 3.3 350 ly

4.2

ν 4.1

Uranus discovered here in 1781

λ 3.6

Alhena γ 1.9 105 ly

ν 4.4

ξ 4.5

ξ 3.4 57 ly

M67

6th-mag. open cluster; seen in binocs; individual stars only in scope
Dia. = 15'
2,500 ly
Width = 13 ly

CANIS MINOR

γ 4.3

β 2.9 170 ly

ORION

μ 4.1

α Procyon
0.4
11.4 ly
Lum. = 6
Dia. = 2

Procyon has white-dwarf companion at 15 AU; period 40 years
Dwarf mass = 65% of Sun

Milky Way

α Betelgeuse

Chart 17 ▼

Chart 16 ▶

CHART 18

GEMINI CANIS MINOR CANCER

This zone, east of Orion, is high in the south from midwinter to early spring.

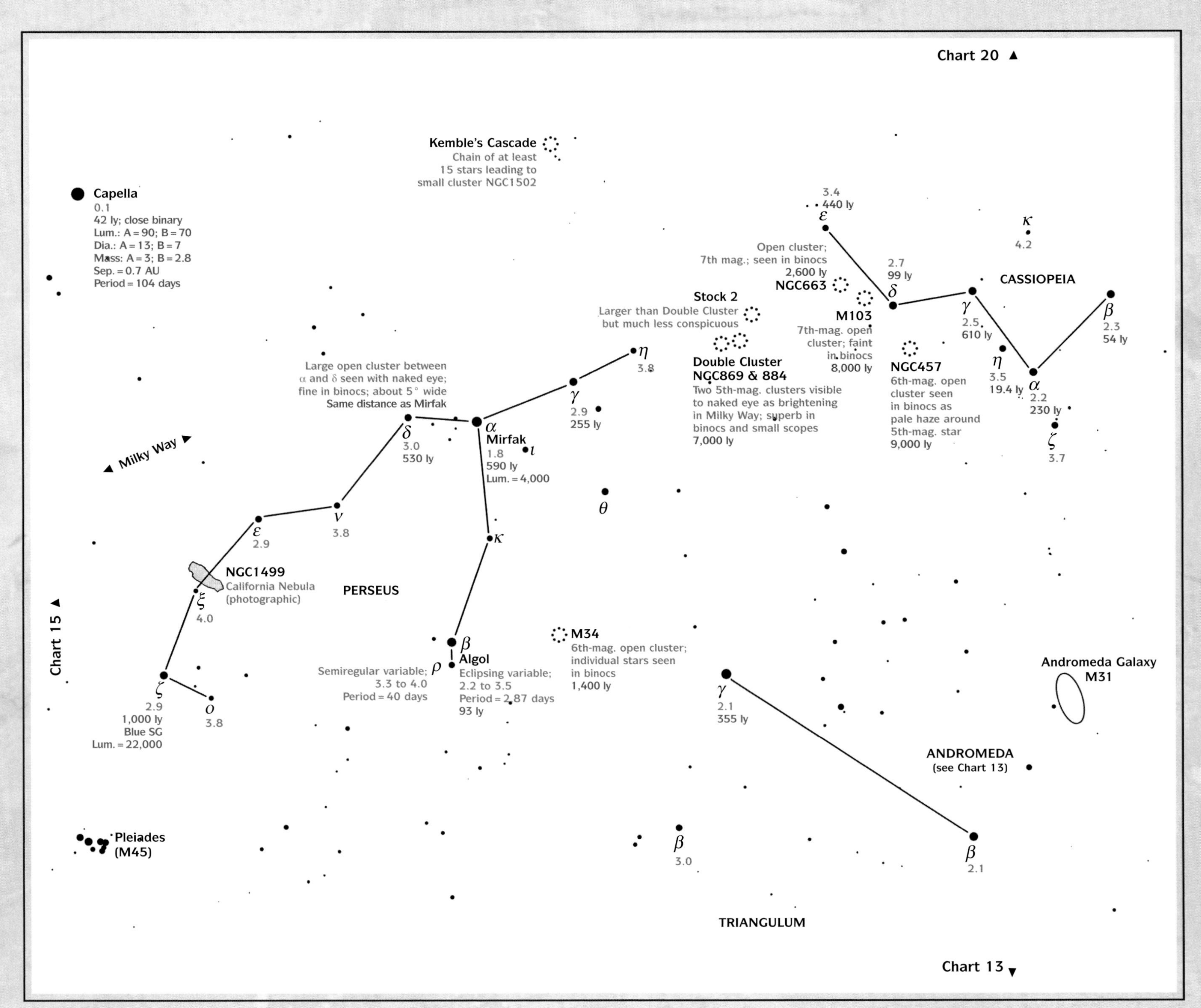

Chart 20
Kemble's Cascade
Chain of at least
15 stars leading to
small cluster NGC1502
Capella
0.1
42 ly; close binary
Lum.: A = 90; B = 70
Dia.: A = 13; B = 7
Mass: A = 3; B = 2.8
Sep. = 0.7 AU
Period = 104 days
3.4
440 ly
Open cluster;
7th mag.; seen in binocs
2,600 ly
NGC663
2.7
99 ly
CASSIOPEIA
κ
4.2
Stock 2
Larger than Double Cluster
but much less conspicuous
M103
7th-mag. open
cluster; faint
in binocs
8,000 ly
2.5
610 ly
β
2.3
54 ly
η
3.8
Double Cluster
NGC869 & 884
Two 5th-mag. clusters visible
to naked eye as brightening
in Milky Way; superb in
binocs and small scopes
7,000 ly
NGC457
6th-mag. open
cluster seen
in binocs as
pale haze around
5th-mag. star
9,000 ly
η
3.5
19.4 ly
α
2.2
230 ly
ζ
3.7
Large open cluster between
α and δ seen with naked eye;
fine in binocs; about 5° wide
Same distance as Mirfak
γ
2.9
255 ly
δ
3.0
530 ly
α
Mirfak
1.8
590 ly
Lum. = 4,000
ι
θ
Milky Way
ε
2.9
ν
3.8
κ
NGC1499
California Nebula
(photographic)
PERSEUS
ξ
4.0
Chart 15
M34
6th-mag. open cluster;
individual stars seen
in binocs
1,400 ly
β
Algol
Eclipsing variable;
2.2 to 3.5
Period = 2.87 days
93 ly
ρ
Semiregular variable;
3.3 to 4.0
Period = 40 days
ζ
2.9
1,000 ly
Blue SG
Lum. = 22,000
ο
3.8
γ
2.1
355 ly
Andromeda Galaxy
M31
ANDROMEDA
(see Chart 13)
Pleiades
(M45)
β
3.0
β
2.1
TRIANGULUM
Chart 13

Chart 19

Perseus Cassiopeia Andromeda

In northeast in autumn, overhead in winter, northwest in spring.

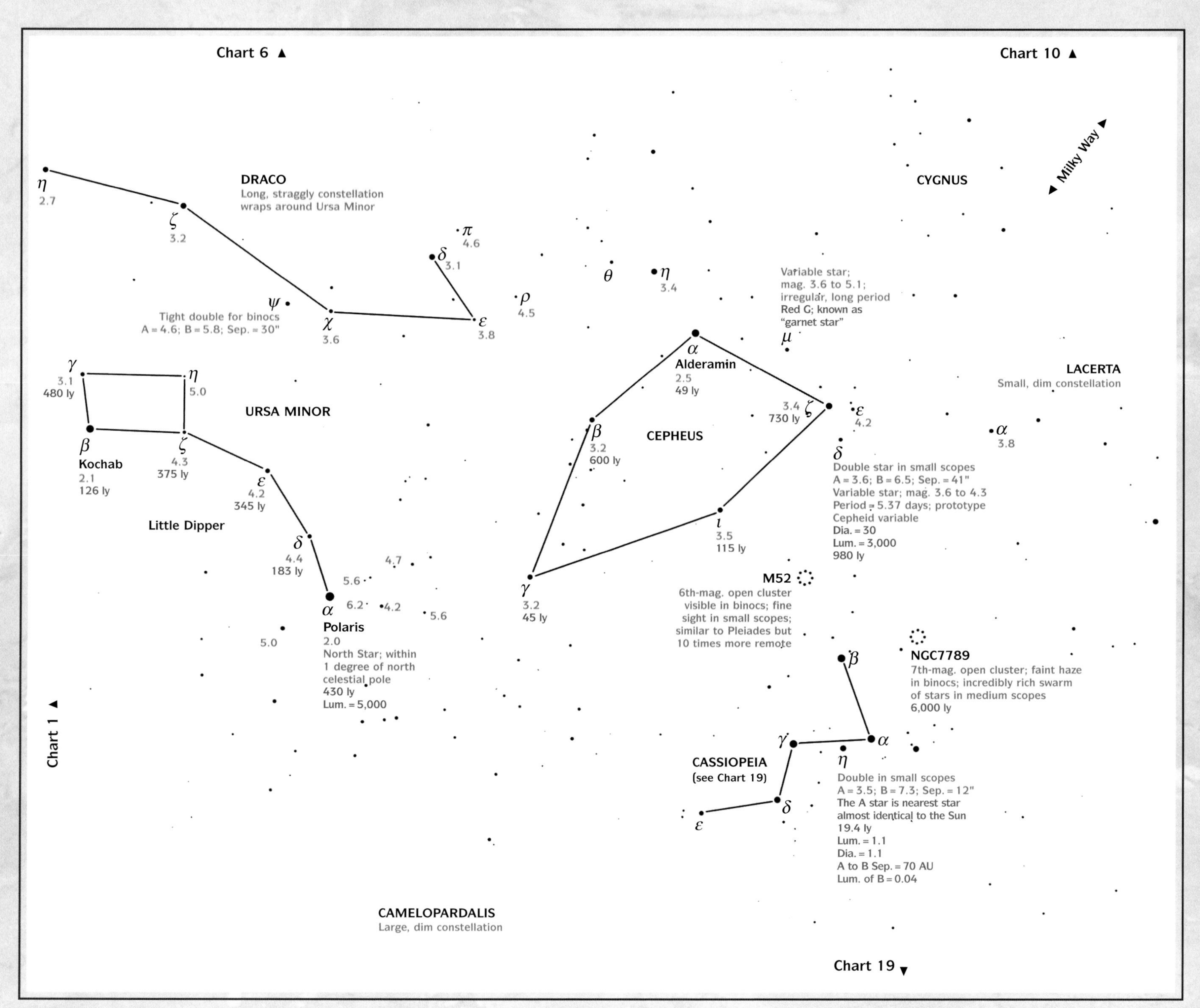

Chart 20

Cepheus Cassiopeia Ursa Minor

All of this chart except bottom third is visible all year; overhead in autumn.

CHAPTER SEVEN

THE PLANETS

Now my own suspicion is that
the universe is not only queerer than we suppose
but queerer than we can suppose.

J.B.S. Haldane

The oldest signs of culture on Earth—sketches on the walls of caves and markings on pieces of bone—reveal that as far back as 30,000 years ago, humans were trying to understand what they saw in the sky. Symbols representing the constellations are found on cuneiform tablets from the first stirrings of civilization in Mesopotamia more than 5,000 years ago. But what intrigued the ancient astronomers the most were five bright wandering stars—the planets—objects that were, to our ancestors, propelled by some magical force.

Masquerading as stars, these celestial rogues roam across the sky in a well-defined belt of constellations known as the zodiac. Today, we know that the planets' orbital geometry dictates that they remain confined to the ecliptic, the plane of the solar system, which has, as its backdrop, the zodiac constellations. Therefore, no planet ever swings by the belt of Orion or wrecks the configuration of the Big Dipper.

All the charts in this book have the ecliptic marked. If some bright object not indicated on the charts is observed in this region of the night sky, it is very likely a planet. Of the eight planets in the solar system besides Earth, five are visible to the unaided eye: Mercury, Venus, Mars, Jupiter and Saturn, all of them as bright as or brighter than first-magnitude stars. Of the remaining three, Uranus and Neptune can be seen in binoculars, and Pluto requires at least a 6-inch telescope.

Mercury, the planet nearest the Sun, is rarely seen, because its tight orbit keeps it so close to the Sun that it pops into view for only a few weeks each year. When it is observed, the planet looks like a yellow zero-magnitude star huddled in the twilight glow after sundown or low in the east just before sunrise. Unless you are specifically looking for Mercury, you'll probably never notice it.

Venus, the second planet out from the Sun and the one that comes nearest Earth, is the sky's premier jewel. At magnitude –4,

Above: Hubble Space Telescope view of Saturn taken in infrared light; arbitrary colors have been applied to the different temperatures detected. Facing page: Five-inch refractor photograph of Jupiter shows dark blemishes caused by the impact of Comet Shoemaker-Levy 9 in 1994.

Venus is so bright that anyone unfamiliar with the planet might think it is not a celestial object at all. Venus, dazzling white, is visible in the early-evening or early-morning sky for several months each year. Because Venus's orbit, like Mercury's, is between the Sun and Earth, the planet is confined to a wedge on either side of the Sun. It can never be seen for more than four hours after sunset or before sunrise.

Mars changes brightness far more than any other planet because its distance from Earth varies by a factor of four, from 0.4 AU to 1.6 AU. Maximum brightness is magnitude –3, but that occurs only at its closest possible approach to Earth (next in 2003). On average, Mars is first or zero magnitude and shines with a distinctive rusty hue, which is caused by sunlight reflecting off its reddish deserts. Mars can move across more than half of the sky during a year, making it the most interesting of all planets to watch as it tracks among the stars.

Jupiter, its magnitude varying from –2 to –3, is brighter than any star but never as bright as Venus. Shining with a steady, creamy white glow, Jupiter is unmistakable. Like Mars and Saturn, its orbit is beyond the Earth's, and consequently, it can, at times, be seen all night stationed anywhere along the ecliptic. Jupiter spends about a year in each zodiac constellation, completing one trip around the Sun in 12 Earth years.

Saturn is the planet most often mistaken for a star, since its brightness equals that of stars such as Regulus, Spica and Antares, rather than exceeding their brightness, as Jupiter and Venus do. And unlike Mars, Saturn does not have a distinctive color, appearing simply as a pale yellow orb. Saturn requires 29½ years to complete its orbit and therefore spends at least two years in each zodiac constellation.

A long, steady look at a planet will usually reveal that it is not twinkling. Planets rarely twinkle like the stars, which invariably flicker, even on the stillest nights, due to atmospheric turbulence. The ubiquitous ripples in the Earth's air blanket easily disturb the pinpoint images of stars to generate the twinkling. Planets, on the other hand, are not pinpoints but tiny disks—too small for the eye to resolve but big enough for their light to be generally unaffected by the ripples, unless the atmosphere is abnormally agitated.

The surest way to identify planets, however, is to know

The Sun and its family of planets are arrayed here to show their comparative sizes. On this scale, the correct distance from the Sun to Earth would be about 30 meters. Neptune and Pluto would be one kilometer away from the soccer-ball-sized Sun.

Is It Possible to See Planets or Stars During the Day?

In general, the answer to this question is no. The only celestial object besides the Sun that is normally visible during the day with the unaided eye is the Moon, and even it is easy to overlook. But for several months each year, the next brightest object, Venus, is dazzling enough to be seen with the unaided eye in daylight.

Before Venus can be located in the daytime sky, a preliminary sighting in a dark evening sky will establish its general location. (Of course, this must be done during a suitable evening-visibility window—see the table at the top of page 134.) Try to catch Venus as soon as the sky darkens after sunset. Once you spot it, maneuver yourself to mark its position with a sight line through the top of a telephone pole, a chimney or anything that projects into the sky. The next clear night, stand in the same location and look for Venus earlier in the evening, slightly above and to the left of the marked position.

Venus appears to move about the width of a thumb held at arm's length (roughly two degrees) in eight minutes. This is not Venus's motion but, rather, the Earth's rotation. Therefore, if Venus was previously observed 15 minutes after sunset, search the sky right at sunset two outstretched thumb widths (roughly four degrees) above and to the left of the marked position. Keep backing up in this manner until Venus is viewed well before sunset. By using this method, it's easy to find Venus in a clear blue sky an hour or more before sunset. However, the sky must be deep blue; any haze greatly reduces the contrast between the planet and the sky and virtually rules out a daylight sighting.

A more direct method of daytime observation is to scan for Venus with binoculars. The planet is surprisingly bright in these instruments, but without some guidelines as to where to look, it might require a fairly lengthy search.

Venus is usually 10 times brighter than the brightest star and 5 times brighter than Jupiter, the next brightest object in the night sky. Although I have never succeeded in picking out Jupiter with the unaided eye before sunset, I have identified it with binoculars around sunset.

Sighting bright stars by day is possible when using telescopes equipped with accurate setting circles. I have found Vega, Sirius, Procyon and Altair this way, but it is really an academic exercise. There is a persistent legend, however, that stars can be seen with the unaided eye in full daylight from the bottom of a dark shaft. The specific example cited in many books is the story that the star Thuban, which

was once the pole star, was visible up a sighting hole constructed within the Great Pyramid of Cheops, from the pharaoh's burial chamber to the outer face. It is said that this star could be seen once each day from the pitch-black cavern within the pyramid.

In 1964, two of these so-called sighting holes—long since filled with rubble and debris—were examined in detail by astronomers and Egyptologists. One hole did indeed point to Thuban, while the other aimed at Alnilam, the middle star in Orion's belt. But because the sighting holes were not straight enough to sight through, the researchers concluded that the shafts were likely symbolic of the pharaoh's celestial journeys to these two stars, each of which had significance in ancient Egyptian religion.

To determine once and for all whether a star could be seen in daylight from the bottom of a dark shaft, a University of Ohio astronomy professor took his students to the base of an abandoned smokestack an hour before the star Vega passed precisely overhead. With their eyes fully dark-adapted in the gloom below, the students waited until the interval when Vega was aligned with the stack's opening. The appointed time came and went. They saw nothing, even though some resorted to using binoculars. The brightness of the sky was too overwhelming.

Above: The crescent Moon and Venus were photographed in the daytime sky with a 200mm telephoto lens. By far the best time to spot Venus in the daytime with the naked eye is when it is within a few degrees of the Moon. Dates of these events can be ascertained from astronomical almanacs and computer planetarium programs. Since the Moon is easy to locate in full daylight, it acts as a visual guidepost to Venus.

Astronomy From the City

Planet observing with binoculars or a telescope is one of the few astronomical activities that can be conducted almost as well from the city as from dark rural locations. The planets are bright enough, except when near the horizon, to cut through city murk and pollution, providing accessible targets for urban astronomers. A telescope set up on a patio provides virtually the same views of Jupiter, Saturn, Venus and Mars as can be obtained at a farm or cottage. In the 1960s, my 7-inch refractor, located in a backyard surrounded by streetlights and immersed in the glow of suburban Toronto, afforded some of the best views of the planets that I have ever had.

Seeing in the city can be just as good (or bad) as elsewhere, and there is plenty of anecdotal evidence that it can even be better from time to time. That's because the air pollution from cities and the heat generated by pavement and buildings can form a microclimate of stable air over an urban area if the atmosphere is calm. Under such conditions (most often experienced on sultry summer evenings), the seeing can be extremely steady, revealing exquisite telescopic detail of the planets.

where they are relative to the constellations or, in the case of Venus and Mercury (which are usually seen before complete darkness), where and when to look. All these facts are supplied in the tables at the end of this chapter. The outer planets Uranus, Neptune and Pluto are below naked-eye visibility. To track these three planets with optical aid, use maps such as those in the *Observer's Handbook* of The Royal Astronomical Society of Canada or the *Astronomical Calendar* (see Resources).

Mercury

The planet nearest the Sun, Mercury is almost a twin of the Earth's Moon. Both sport heavily cratered surfaces that have remained essentially unaltered for the last three-quarters of the 4.7 billion years since the solar system's birth. The craters are scars from what must have been ferocious bombardments of debris left over from the formation of the planets. Although the Earth's crust has undergone many changes since then, the surfaces of Mercury and the Moon have not.

It is as difficult to observe details on Mercury as it would be to examine the surface of the Moon if it were 300 times farther from Earth. A telescope magnifying 150 times (typical for planetary inspection) shows Mercury just half the size of the Moon seen with the unaided eye. Added to that is Mercury's location near the Sun's glare. For these reasons, only a few

Left: Mercury can be as bright as magnitude –1.5, but it always hugs the morning or evening horizon, as seen in this photograph. Above: The two brightest objects seen over this early-morning cityscape are Venus and Mercury.

Times when these two planets are visible in the morning and evening sky are given in the tables at the end of this chapter.

smudges have ever been recorded from Earth. Our knowledge of Mercury's surface stems entirely from photographs taken by the U.S. spacecraft Mariner 10, which flew past Mercury in 1974, the sole human artifact ever to visit that planet.

However, backyard astronomers can see the phases of Mercury. All but the smallest telescopes reveal a tiny crescent or half phase during the latter half of Mercury's prime evening-visibility window each spring. The phases are simply the varying amounts of the day and night sides visible from Earth as Mercury swings in its orbit.

Because Mercury is easily identified only when it is close to the western horizon after sunset (or the eastern horizon before sunrise), horizon-induced poor seeing often causes a mushy telescopic view. Mercury, a small, scorched, dead world on the solar system's inner fringe, is a challenge to the backyard astronomer. I find satisfaction in just locating the elusive little planet, and I feel fortunate if I can get one sharp telescopic peek at Mercury each year.

Venus

For five or six months every year and a half, a brilliant object hovers over the western horizon each clear evening after sunset. This is Venus, second only to the Moon in brightness in the night sky. Venus's nearest rival, Jupiter, is never more than half as radiant, while Sirius, the brightest star in the night sky, is only 8 percent as luminous. Venus's dominance is largely due to its nearness to Earth. No other planet approaches as closely. Its minimum distance is 0.3 AU, about 100 times the Moon's distance.

As the unsurpassed beacon in the night sky, Venus has figured prominently in the religions and cultures of many peoples. But none ever elevated Venus to the level of importance it held for the Mayan civilization of Mexico. At their peak, about 1,200 years ago, the Maya developed a sophisticated calendar system based on a 584-day period—the interval required for Venus to return to the same position in the sky. They called Venus the Ancient Star and were obsessed with its cycle of visibility.

Spanish historian Bernardino de Sahagún studied the Maya in the 16th century and reported that enemies were sacrificed when Venus became visible after being lost for a few weeks in the Sun's glare. "When Venus made its appearance in the east," he wrote, "[the Maya] sacrificed captives in its honor, offering blood, flipping it with their fingers toward the planet."

Another reason for Venus's dazzling appearance is a cloak of brilliant white clouds that reflects 72 percent of the Sun's light back into space. At only two-thirds the Earth's distance from the Sun, Venus receives twice as much sunlight.

Venus is regarded by many scientists as an Earth gone wrong. The planet is almost identical in size and mass to Earth, yet it is shrouded in an atmospheric cloak 90 times as dense. This blanket has turned Venus into a planetary greenhouse,

effectively trapping the solar radiation that penetrates the clouds. The surface temperature from pole to pole is within a few degrees of 460 degrees C.

The Venusian atmosphere is almost entirely carbon dioxide laced with sulfuric-acid droplets. It is inconceivable that any form of life as we know it could have evolved on Venus under these conditions. Even exotic science fiction scenarios for silicon-based and other life forms seem unrealistic in the Venusian environment.

I like to observe Venus telescopically just around sunset, when it looks like a cue ball suspended in the deepening blue. The snow-white surface—the cloud blanket—is totally featureless, its blank stare an appropriate mask for the furnace-hot wasteland below. Like Mercury, Venus cycles through phases from nearly full to a thin crescent, but Venus's phases are much

Above: The relative motion of the Moon, Venus and Jupiter over 24 hours is shown in this pair of images taken about 5:30 a.m. on two consecutive mornings, April 22 and 23, 1998. Venus is the brighter planet. Left: Telescopic view of Venus.

easier to detect than Mercury's. The crescent can be seen in steadily held binoculars, and a good telescope will produce a sharply defined image of the planet, large and dramatic when Venus nears its closest point to Earth once every 1½ years. At such times, it is a beautiful sickle-shaped crescent.

In general, though, Venus is not a compelling telescopic subject, since nothing apart from the phases can be seen. When it decorates the evening sky, Venus is usually the first object I turn my telescope toward. But after a few minutes, I begin tracking down more varied targets. The real face of the queen of the night is never unveiled to backyard astronomers.

Mars

The old Mars, as we knew it before the space age, was a wonderful world of heroes, maidens, bizarre creatures and, most captivating of all, a dying civilization desperately attempting to prolong its existence by constructing a global network of canals to preserve dwindling water supplies. Peering from Earth, turn-of-the-century astronomers saw those canals—or thought they did.

Today, the mystery of the famous canals is gone. They proved to be not waterways but optical illusions born in the minds of observers who unconsciously linked subtle detail near the threshold of vision into linear features. Today, speculation about Martian civilizations has given way to the real Mars, a planet midway between Earth and the Moon in size and surface conditions. Spacecraft images of deserts, craters and colossal volcanoes that would dwarf Mount Everest have revealed a world less like the pre-space-age picture than anyone suspected. A major turning point was the negative results from the sophisticated life-searching equipment on board the American Viking 1 lander. The Mars of our dreams was gone for good.

I vividly remember the final transition from the old Mars to the new. It was a warm summer evening in July 1976 when the first Viking touched down on Mars. I was at mission control at the Jet Propulsion Laboratory in Pasadena, California, with about 200 scientists and an equal number of journalists and science fiction writers anxiously awaiting the first surface photographs to be beamed from Mars. The images began to build up line by line on the television monitors, and a boulder-strewn landscape of sand dunes slowly emerged. Science fiction writer Ray Bradbury, who happened to be standing beside me, said quietly: "From this moment on, we don't have to imagine what Mars is like anymore."

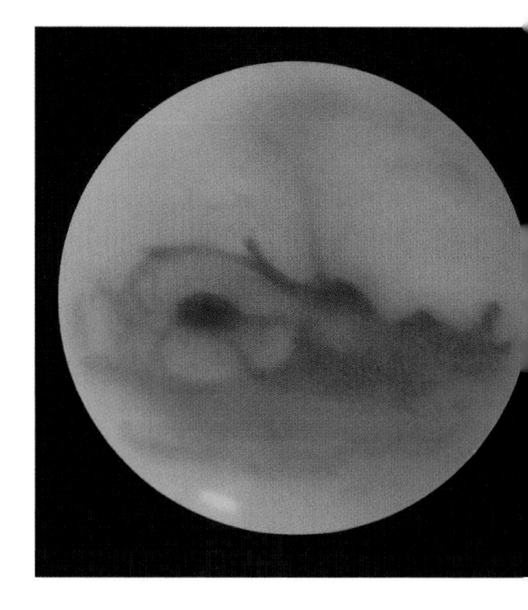

And so it has been with most of the other major worlds in the solar system. They have been removed from the borderline of science and science fiction and brought sharply into focus by the electronic eyes of robot spacecraft that Earthlings have flung to all but the outermost planet of the solar system. Paralleling this, the focus of backyard astronomy has changed. Prior to the space age, planetary observation was the chief activity of amateur astronomers, because so little was known about our neighboring worlds. The aura of the unknown was a powerful magnet that kept hobbyists hunched over their eyepieces for hours to catch a glimpse of a Martian canal or an alteration in the red planet's dark zones (believed to be vegetation as recently as 1965 by a few experts).

Today, the emphasis has shifted. Examination of Mars through the telescope is more casual, less encumbered by in-

Sketch of Mars, top, was made by the author in September 1988 using a 7-inch apochromatic refractor. In 1997, Earth was treated to spectacular Martian panoramas beamed back by the Pathfinder spacecraft and its wheeled companion Sojourner.

quiry—more like a tourist's view of a well-known vacation island from an aircraft. Interesting, somewhat exotic, but not as mysterious. When I view the Martian disk now, I think of the winds that whip the dunes of the vast deserts and howl down the great Mariner Valley, a canyon five times deeper and a hundred times longer than the Grand Canyon. I see from afar the endless plains painted red by layers of iron-oxide minerals, which give the planet its unique rusty-colored hue, and the vast polar ice caps trimmed by a sheet of solidified carbon dioxide—part of the planet's thin atmosphere, frozen to the ground over the winter.

Yet, as it has for millennia, the red "star" dominating the evening sky still attracts attention. Every 26 months or so, Earth catches up to Mars and overtakes it. The point of passing, or closest approach, is called opposition. Because Mars has an elliptical orbit, the distance between the two planets can vary from 0.37 AU to 0.68 AU from one opposition to another. However, even the greater distance is considered close by planetary standards. Here are a few recent and upcoming oppositions: March 17, 1997 (0.66 AU); April 24, 1999 (0.58 AU); June 13, 2001 (0.45 AU); August 28, 2003 (0.37 AU); November 7, 2005 (0.46 AU); December 24, 2007 (0.59 AU); January 29, 2010 (0.66 AU).

The problem for backyard astronomers is that Mars is a small planet, only half the size of Earth. Even at its minimum distance, Mars appears barely larger than the globe of Saturn and nowhere near as big as Jupiter. Sometimes, when Mars and Earth swing to opposite sides of their orbits, the red planet shrinks to the same apparent size as Uranus. So it is not surprising that Mars is, as often as not, just a pale coral-hued dot in the telescope's field of view.

Within a few weeks of opposition, however, Mars does reveal itself to the backyard observer. Because of its tenuous atmosphere (0.7 percent as dense as the Earth's), Mars is the only planet whose surface features are clearly visible from Earth. The most prominent markings are the major dark areas Syrtis Major and Acidalium and one or both of the brilliant white polar caps.

A complete rotation of Mars was captured by the Hubble Space Telescope in 1997 while the planet was at its closest to Earth. Syrtis Major, the most prominent dark feature in small telescopes, is just below center in the fourth image.

My first good look at Mars was through a high-quality 3-inch refractor, but I was initially disappointed to see nothing more than the white button of one of the polar caps. It was weeks before I was able to train my eye to discern more detail. Even now, when Mars returns every other year, I have to educate my eye again, although it takes only a night or two. Telescope, mind and eye work together to reveal more and more each night. Since Mars' rotation is only 40 minutes slower than the Earth's, the same face is seen 40 minutes later the following evening, which aids in the familiarization process. Phobos and Deimos, the two moons of Mars, are much too small for detection in all but the largest amateur telescopes.

During each opposition, I try to make enough sketches of Mars to produce a small map of the planet. I use a two-inch circle and a soft pencil and keep detailed notes of the intensity of the features, indicating whether they were easy or difficult to see. I record the time and the date so that I can calculate the planet's central meridian longitude from tables in the *Observer's Handbook* (see Resources). Then I can plot what I observe.

The thrill of seeing Syrtis Major or of catching a fleeting glimpse of the Chryse Plain, where Viking 1 landed, is the reward. Knowing that the great Olympus Mons volcano looms between Tharis and Amazonis stokes the imagination. The new knowledge of Mars revealed by the television eyes of Viking, Pathfinder and its follow-on space probes has shown the red planet to be an exciting, varied world that had remained unknown for centuries of Earth-bound investigations.

The Asteroid Belt

The popularized image of the asteroid belt as an outer-space pinball machine, with craggy crater-pitted boulders practically bumping into each other, makes exciting science fiction but departs from reality. The zone occupied by the asteroid belt, between the orbits of Mars and Jupiter, is enormous. Even though more than a billion asteroids the size of a house or larger roam in this area, the space they occupy is so vast that an astronaut standing on one of the rocky bodies would only rarely see another asteroid appearing brighter than a third- or fourth-magnitude star.

Asteroids were unknown before the 19th century—but not entirely unsuspected. Astronomers were puzzled by the wide, apparently vacant zone between the orbits of Mars and Jupiter, which spoiled the otherwise regular progression of worlds outward from the Sun. So strong was the feeling that something must occupy this region that several astronomers spent years searching for the "missing" planet.

On the last evening of the 18th century, Giuseppe Piazzi, an Italian monk who was revising a star atlas, noticed a star that was not plotted. He checked its position the following night and found that it had shifted slightly. When the new body's orbit was calculated, astronomers realized that it wheeled around the Sun right in the middle of the Mars-Jupiter gap. The object, named Ceres, was apparently much smaller than the Moon.

Skywatchers never had a chance to get used to the idea of one small planet between Mars and Jupiter because a second one, Pallas, was soon discovered. Two more tiny planets, Juno and Vesta, were tracked down during the following years.

Today, more than 7,000 asteroids have been observed long enough for a determination of their precise orbital paths. The number of asteroids one kilometer in diameter or larger is estimated to be half a million. Ceres, the biggest by far at 930 kilometers in diameter, is one-third of the estimated mass of all the rest of the asteroids combined. Next are Pallas (530 kilometers), Vesta (500 kilometers) and Hygeia (430 kilometers).

Yet of all these asteroids, only Vesta is occasionally barely

Jupiter and its four largest moons rank among the night sky's top telescopic targets. This full-color image is a montage of five separate images from the Galileo spacecraft that orbited and studied Jupiter during the 1990s. The moons are, from top to bottom, Io, Europa, Ganymede and Callisto. Inset is a small-telescope view of Jupiter and the same four moons.

visible to the unaided eye. However, dozens of asteroids can be seen as starlike objects in backyard telescopes. Sky charts of their positions are published in the *Observer's Handbook*, the *Astronomical Calendar* and *Sky & Telescope* magazine (see Resources).

The existence of an asteroid belt instead of a planet could mean one of two things: that a planet was unable to form and the debris is evidence of the aborted process or that a planet did form but somehow shattered. The exploded-planet hypothesis has relatively few adherents. Most astronomers currently support the theory which proposes that an original family of smaller objects never coagulated into a major planet due to the disruptive gravitational influence of nearby Jupiter, the most massive planet in the solar system.

Jupiter

In one of Arthur C. Clarke's science fiction novels, an interplanetary spaceship approaches Jupiter from just beyond the orbit of Callisto, outermost of the planet's four big satellites. Here, more than two million kilometers from the solar system's largest planet, one of the crew gazes out the ship's window toward the colossal globe of Jupiter, which resembles a multicolored beachball suspended in the sky by an invisible string.

As the space traveler watches, mesmerized by the quilt of storm-riddled clouds swirling across the face of Jupiter, he notices that the great globe is spinning rapidly. Clouds that only a few hours ago were at the center of the planet are now moving out of sight as new ones sweep into view. Clarke goes on to explain that the planet is not a world of rock or ice but is a ball of liquid and gaseous hydrogen, the largest of the solar system's four gas giants.

What intrigues me about Clarke's description of the view is that it's not much different from what I have seen when examining Jupiter through my 6-inch telescope with an eyepiece magnifying about 180 times. A cooperative atmosphere that allows a distortion-free view of Jupiter puts me, in effect, less than a million kilometers beyond Callisto. It is the next best thing to the view that I would have gazing out the window of Clarke's interplanetary spaceship.

Binoculars are all you need to begin observing the Jovian system. They will reveal the movements of the four largest moons that orbit Jupiter in periods ranging from just under 2 days for Io (the nearest of the big four) to about 17 days for Callisto. Steady the binoculars by resting them on a railing or a fence, since your arms alone cannot hold them still enough for astronomical observations. For the best results, support them rigidly by using an adapter that clamps the binoculars to a camera tripod.

As the moons of Jupiter shuttle back and forth from side to side, they provide an ever-changing solar system in miniature. Occasionally, the satellites are paired on either side of the belted gas planet or are all lined up on one side. Some of the moons may be lost in Jupiter's glare, or they may be behind or in front of the planet. Even in observatory telescopes, only the four largest of Jupiter's 16 moons are visible. The other satellites, all less than 300 kilometers in diameter, are photographic targets.

The major satellites in order from Jupiter are Io, Europa, Ganymede and Callisto. At magnitude 4.6, Ganymede, the largest, is theoretically bright enough to be visible without optical aid. However, Jupiter's brilliance prevents such observation—at least that has been the accepted wisdom on the subject. Yet for many years, there have been persistent reports by people who claim to have seen Ganymede with the naked eye. Some say that they have also detected Callisto, the most distant of the large moons, even though it is only magnitude 5.6, near the threshold of vision. I cannot see either of them myself with the unaided eye, and I have always concluded that these sightings are biased because the observers know the satellites are there—a mind-eye combination which gives the illusion of a small dot beside the brilliant luminary.

A sketch of Jupiter as seen through a 3-inch refractor shows the planet's cloud belts and, looking like a bullet hole in the giant planet, the inky shadow of one of its four large moons.

After many years of inconclusive debate on this subject, it appears that at least some of these sightings are accurate. According to Xi Ze-Zong of the Chinese Academy of Science, Gan De, one of China's earliest astronomers, left records of observations of one of Jupiter's moons in 364 B.C. Gan De wrote: "Jupiter was very large and bright; apparently, there was a small reddish star appended to its side." This passage, says Xi Ze-Zong, almost certainly means that Gan De saw a Jovian satellite.

To test the validity of Gan De's statement, an experiment was conducted in a Chinese planetarium which revealed that people with good eyesight can see a satellite the same apparent brightness as that of Ganymede and Callisto at their farthest orbital positions from Jupiter. The tests also showed that it would be even easier to see a blended image of the two. It would seem that Galileo's discovery of Jupiter's moons in 1610 was preempted by almost 2,000 years.

Tracking a Jovian moon as it casts its shadow on the disk of the giant planet is one of the great telescopic sights in astronomy. This phenomenon is visible in any decent backyard telescope at about 100x and is especially prominent when the tiny ink-black shadow of a moon can be spotted on one of the brighter zones of Jupiter's clouds.

Because the satellites range in size from Europa, slightly smaller than our Moon, to Ganymede, 1½ times the Moon's diameter, and because of their varying distances from Jupiter, their shadows are of different sizes. Io and Ganymede produce slightly more obvious shadows than do Europa and Callisto. Experienced observers track the shadows of all four with 3-inch refractors, but for general conditions and the average eye, a 4-inch or larger telescope is preferred.

Both the satellites' orbital positions and the times when their shadows are visible on the planet's disk are given in the *Observer's Handbook* (see Resources). Also given are the times when each moon dips into Jupiter's shadow or disappears behind the planet. The moons' orbits are exactly in the plane of Jupiter's equator. But since Jupiter stands up almost vertically, with its axis tipped only three degrees, the plane of the equator and therefore the orbits of the satellites are seen nearly edge-on. It's like watching a billiard game with your eyes at table level. The moons swing back and forth without deviating from a thin zone on either side of Jupiter.

The windswept cloud deck of Jupiter is continually changing, the vast dark belts merging with one another or fading to insignificance. The bright zones—actually smeared bands of ammonia clouds—vary in intensity and are frequently carved up with dark rifts or loops, called festoons. The most dramatic action occurs in the equatorial zone. The clouds at Jupiter's equator rotate five minutes faster than those on the rest of the planet: 9 hours 50 minutes compared with 9 hours 55 minutes. This means a constant atmospheric-current interaction as one region slips by the other at approximately 400 kilometers per hour. Besides these disturbances, there are changes in the intensity of the various belts and zones from year to year.

Despite its great distance, Jupiter appears far larger in the telescope than any other planet. Even when seen at their best, all the other planets combined do not equal Jupiter's visible surface area. At least two of the equatorial cloud belts will be seen in any decent telescope, and a quality instrument of moderate size (4-to-8-inch) will reveal a multicolored globe of atmospheric zones.

Jupiter's rapid rotation has caused the great globe to become markedly oval so that it appears to be about 7 percent squashed at the poles. The variety of its cloud features and the choreography of its four large moons make Jupiter one of the top backyard attractions for amateur astronomers.

Saturn

Saturn is the superstar of the night sky, a dazzling showpiece unmatched by anything else visible in the telescope. No one will forget that first telescopic view of Saturn—the chilling beauty of the small pale orb and the delicate encircling rings magically floating in a field of black velvet. Although we are gazing across more than a billion kilometers of space, even the smallest telescope will reveal the rings. The image of Saturn in a quality moderate-aperture telescope can be truly breathtaking.

The rings of Saturn are enormous in extent. From one edge to the other, they measure more than two-thirds the distance from Earth to the Moon. They consist of trillions of tiny ice moonlets whirling about Saturn, each in its own orbit, in periods ranging from 4 hours at the inner edge to 14 hours at the outer edge. In the densest sectors, the moonlets—from particles the size of a speck of dust to boulders the size of a house—form a dazzling celestial blizzard around the planet. Yet if an astronaut were within the rings, there would be little

Above: This section of a page from an amateur astronomer's notebook includes a sketch of Jupiter. Typical backyard telescopes can reveal astounding detail on the giant planet that is challenging to record. Jupiter's cloudy surface changes from year to year in largely unpredictable ways. Facing page: The Voyager 1 space probe captured this image as it approached Jupiter in March 1979.

danger as long as he or she was orbiting Saturn at the same velocity. The debris would then move in stately procession, with relatively few collisions.

From a distance, the rings appear as a solid sheet, precisely aligned by gravitational forces into a zone less than a kilometer thick exactly above Saturn's equator. In apparent size, the rings are about as wide as Jupiter's disk. Saturn itself appears as a pale yellow globe roughly one-third the width of Jupiter. Sometimes, one or two dusky bands are seen at the planet's equatorial region. Findings from spacecraft missions to Saturn have revealed superhurricane wind belts similar to Jupiter's, but an upper-atmosphere ice-crystal haze results in a subdued surface as viewed from Earth. Saturn rotates on its axis in 10 hours 40 minutes, but the planet's bland face makes the spin imperceptible in backyard telescopes.

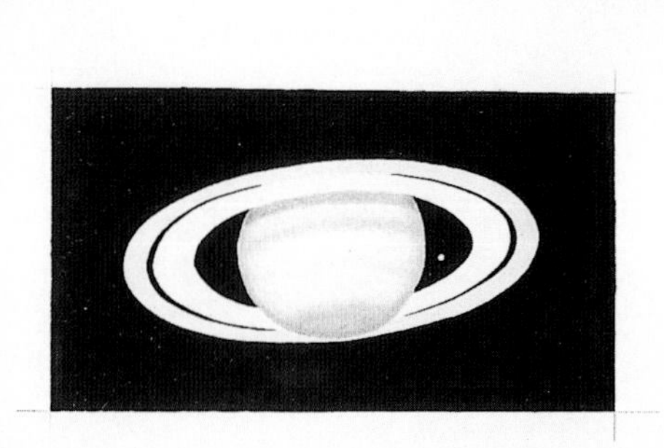
July 2-3, 1989
05:00 - 09:30 UT
Coronation Park, Edmonton
C14 at 148 x , 4" Zeiss at 111 x
Conditions: seeing was excellent from 06:30 to 08:00, may have been the best seeing conditions I've seen.
Observed Saturn pass in front of 28 Sag. (Sky & Telescope says this may be a "once in a life-time" event.) Drawing depicts event at 08:37 UT. Occultation of Crepe Ring was spectacular. Observed star shining through Cassini's and Encke's division. Also observed Neptune, Uranus and Vesta through 7x50 binoculars.

The rings of Saturn are more reflective than the planet's clouds and are therefore noticeably brighter. Gaps and brightness differences define several distinct rings, the two brightest of which are plainly visible in backyard telescopes. They are separated from each other by Cassini's division, a gap about as wide as North America that can be seen with a 3-inch refractor under good conditions. Often easier to detect is the shadow of Saturn on the rings as they curve behind the planet. Also look for the rings' shadow on the disk of Saturn itself.

Saturn controls a family of at least 18 moons. Titan, the largest, is slightly bigger than the planet Mercury and is cloaked in a thick nitrogen blanket denser than the Earth's atmosphere. Titan could be regarded as a small planet orbiting a large planet. The big moon is visible in any telescope as an eighth-magnitude object looping around Saturn in 16 days. At its maximum distance east or west of Saturn, Titan appears about five ring diameters from the planet. Like all the inner satellites in Saturn's system, Titan orbits exactly above the planet's equator. Of the remaining moons, Rhea, at 10th magnitude, is the only one readily seen in telescopes less than 5 inches in aperture. Its orbit lies well within that of Titan, less than two ring diameters from Saturn. Two other moons, Dione and Tethys, are targets for 6-inch and larger telescopes. They are often nestled in the glare just beyond the rings, but finding them is one of the bonuses in Saturn observing. Try using averted vision, as described on page 92.

From one year to the next, Saturn's rings noticeably vary their inclination, ranging from edge-on in 1995 and 2009 to a maximum tilt of 27 degrees in 2002. Saturn is the most oblate planet in the solar system, bulging at the equator by an amount equal to the Earth's diameter. This difference is obvious when the rings are within a year or two of being edge-on.

Left: A notebook entry records the extremely rare passage of Saturn in front of the fifth-magnitude star 28 Sagittarii in 1989. Above: This superb drawing of Saturn by planetary observer Paul Doherty shows the view in exceptionally good seeing with a 15-inch telescope. This approaches the maximum amount of detail visible under the most favorable circumstances in the best amateur telescopes.

The Outer Planets

Beyond Saturn, the solar system becomes bleak. Uranus, Neptune and Pluto are not only remote but also dim due to their residence in the gloom so far from the Sun. Uranus (pronounced YER-an-us, not your-AY-nus, as you were probably taught in school) is technically at the threshold of vision, at magnitude 5.8. For a proper identification, however, you'll need binoculars and a current guide map from one of the astronomy magazines or the *Observer's Handbook* (see Resources). I have had no trouble spotting it from a city-apartment balcony using 7x50 binoculars. But seeing something more than a starlike dot is another matter. A good telescope and a power of 100 or more are needed to make the little bluish green disk obviously nonstellar. Even its largest moons (less than half the size of Jupiter's) are invisible in all but the most sophisticated amateur equipment. Unlike Saturn, which gains more than a magnitude in brightness when its rings are tipped toward us, Uranus benefits not a smidgen from its necklace of narrow black rings. In fact, Uranus's rings are so dim, they have never been visually detected from Earth (only electronically amplified images reveal them).

Uranus is so inconspicuous that it was mistaken for a star dozens of times before its accidental discovery in 1781 by English amateur astronomer William Herschel using a 6-inch Newtonian reflector. Since that time, various observers using large telescopes have reported pale belts parallel to the equator. Theoretically, it is possible to see detail on Uranus; at 450x, the planet appears the same size as the Earth's Moon does to the unaided eye. But Hubble Space Telescope images of Uranus reveal only faint smudges on a uniform aquamarine smog deck that conceals whatever cloud belts exist deeper in the atmosphere.

Neptune, half of Uranus's apparent size, is an even tougher object to discern as nonstellar in a telescope. Finding it is enough reward for most backyard astronomers. At magnitude 7.7, it is also a binocular object but is more difficult to identify.

Just seeing Pluto is a major achievement that few amateur astronomers can claim. It is so dim, at magnitude 13.7, that a 6-inch telescope is minimum equipment. A larger instrument makes the task easier but by no means simple. Use the charts in the *Observer's Handbook* or *Sky & Telescope* magazine (see Resources). To confirm a sighting, two observations are necessary to reveal Pluto's motion among the stars. Otherwise, Pluto cannot be distinguished from a star.

A rare close grouping of three planets in mid-June 1990 is remembered by those who saw it as one of the most striking conjunctions of a lifetime. The planets are, in order of brightness, Venus, Jupiter and Mars.

Visibility of the Planets 1998-2010

The Planets: Naked-Eye Appearance

Venus, sometimes called the morning or the evening star, is the brightest planet in our solar system. Its dazzling diamond-white appearance and substantially greater brilliance than any other star-like object make it easily recognizable.

Jupiter is the second brightest planet, but it still outshines all the fixed stars. It shines with a slightly off-white creamy hue. Because it takes 12 years to orbit the Sun, Jupiter spends about a year in each constellation of the zodiac.

Saturn, similar in color to Jupiter, is the same brightness as the brightest stars. The slowest-moving naked-eye planet, Saturn takes 29½ years to orbit the Sun and spends about two years in each constellation of the zodiac.

Mars shines with a definite ocher or pale rusty hue. It varies greatly in brilliance, ranging from the brightness of the Big Dipper stars to that of Jupiter. At times, it moves rapidly across the heavens, spending as little as one month in a zodiac constellation.

Mercury is the most elusive of the naked-eye planets, being visible only near the western horizon at dusk or the eastern horizon at dawn for two-week intervals a couple of times during the year.

Venus: Periods of Prominent Visibility

Western Sky at Dusk	Eastern Sky at Dawn
late Dec. 1998 to early Aug. 1999	late Aug. 1999 to mid-Apr. 2000
mid-Sept. 2000 to mid-Mar. 2001	late Apr. 2001 to mid-Nov. 2001
early Mar. 2002 to late Sept. 2002	mid-Nov. 2002 to mid-Apr. 2003
early Nov. 2003 to late May 2004	late June 2004 to mid-Jan. 2005
early June 2005 to early Jan. 2006	late Jan. 2006 to mid-Sept. 2006
mid-Dec. 2006 to mid-July 2007	early Sept. 2007 to late Feb. 2008
late Sept. 2008 to mid-Mar. 2009	early Apr. 2009 to mid-Nov. 2009
early Mar. 2010 to mid-Sept. 2010	mid-Nov. 2010 to mid-Mar. 2011

Note: Although Venus is the brightest object in the night sky apart from the Moon, it is often close to the horizon. If possible, observe from a location with an unobstructed horizon in the specified direction.

Prominent Planetary Conjunctions

The Moon and the brighter planets are the night sky's most conspicuous celestial objects, especially when they have close encounters with each other—events called conjunctions. The tabulation below includes the best of, though far from all, the conjunctions visible from North America through 2010. Asterisked items are either particularly impressive or relatively rare. Note especially the April/May 2002 event.

February 23, 1999*	Venus 0.3° from Jupiter in evening sky; spectacular
March 19, 1999	Venus 2° from Saturn; crescent Moon nearby
February 2, 2000	Crescent Moon 1° from Venus in morning twilight
early April 2000	Jupiter, Saturn and Mars cluster low in evening sky; crescent Moon joins them April 6
July 15, 2001	Venus and Saturn less than 1° apart in early-morning sky
July 17, 2001	Crescent Moon, Venus and Saturn form a 3° triangle in morning twilight
August 6, 2001	Venus 1.3° from Jupiter in early-morning sky
November 6, 2001	Venus 0.8° from Mercury in morning twilight
February 22, 2002	Gibbous Moon 0.3° from Jupiter around midnight
April 22 to May 13, 2002*	For the first time in more than a generation (since May 1980), all five naked-eye planets are lined up above the western horizon at dusk; crescent Moon joins the group on May 13; best alignment will be late April, with planets in this order from the horizon up: Mercury, Venus, Mars, Saturn and Jupiter; star Aldebaran near Saturn
May 14, 2002	Crescent Moon 1° from Venus low in west at dusk
June 3, 2002	Venus and Jupiter 2° apart low in west at dusk
June 14, 2002	Crescent Moon less than 2° from Venus (3° in western North America)
November 5, 2004*	Venus and Jupiter 0.6° apart in the east in morning twilight
December 7, 2004*	Crescent Moon passes directly in front of Jupiter from about 4 to 5 a.m., EST (not visible from western half of North America)
June 25, 2005	Venus, Mercury and Saturn cluster within 1.5° of each other in west at dusk
June 27, 2005*	Venus just 0.1° from Mercury in west at dusk; closest encounter of these two planets visible from North America since 1965
June 17, 2006	Saturn and Mars 0.5° apart and just 1° from Beehive star cluster in evening sky
May 19, 2007*	Crescent Moon just 1° from Venus in evening sky; very impressive
July 1, 2007	Venus 0.8° from Saturn low in west in evening twilight
February 1, 2008	Venus and Jupiter 0.6° apart in east in morning twilight
February 4, 2008	Venus, Jupiter and crescent Moon cluster in morning sky before dawn
December 1, 2008*	Venus, Jupiter and crescent Moon cluster in a 3° triangle at dusk
December 31, 2008	Jupiter and Mercury 1.2° apart low in west at dusk
February 27, 2009	Crescent Moon less than 2° from Venus in evening sky
October 13, 2009	Venus and Saturn 0.5° apart in early-morning sky

Where to Find Jupiter and Saturn

	Jupiter	Saturn
1998	In Aquarius in Jan.; too close to Sun in Feb. and Mar.; rest of year on Aquarius-Pisces border	In Pisces until late Mar., when twilight interferes; in Aries after late May
1999	In Pisces Jan. and Feb.; near Sun mid-Mar. to mid-May; in Aries remainder of year	In Aries all year; Sun's glare rules out observation from Apr. to early June
2000	In Aries to mid-Apr.; Sun's glare prevents observation to late June; in Taurus thereafter	Jupiter and Saturn less than 15° from each other all year
2001	In Taurus until solar glare scoops it up in May; seen in Gemini in morning sky after late July	Saturn less than 15° to right of Jupiter until solar glare interferes from late Apr. to early July; then look near Aldebaran
2002	In Gemini until solar glare interferes in early June; seen in Cancer from mid-Aug. to year's end	Near Taurus-Gemini border all year; solar glare prevents viewing from early May to early July
2003	In Cancer to mid-July; emerges from solar glare in Leo in mid-Sept. and remains there into 2004	Visible in Gemini except for late May to late July, when lost in twilight glow
2004	Look for Jupiter in Leo until mid-Aug., when it is lost in twilight; in Virgo in morning sky after mid-Oct.	Remains in Gemini; not visible from early June to early Aug. because of solar glare
2005	In Virgo until mid-Sept., when lost in twilight; in Libra after mid-Nov.	In Gemini until twilight interferes in mid-June; in Cancer after late Aug.
2006	Near Virgo-Libra border for entire year except after mid-Oct., when Sun's glare prevents convenient viewing	In Cancer until lost in solar glare from early July to early Sept.; in Leo thereafter
2007	In Scorpius all year; too close to Sun for observing after mid-Nov.	In Leo to right of Regulus until twilight interferes from mid-July to mid-Sept.; then look to left of Regulus
2008	In Sagittarius all year; Sun's glare prevents viewing in Jan.	In Leo all year; Sun's glow interferes from late July to early Oct.
2009	Lost in twilight glow through to early Mar.; in Capricornus rest of year	Near Virgo-Leo border; solar glare interferes from mid-Aug. to mid-Oct.
2010	In Capricornus in Jan.; twilight interferes with viewing through to early Apr.; in Aquarius thereafter	In Virgo all year; twilight glow prevents viewing from late Aug. to late Oct.

Best Times for Sighting Mercury

	Evening Sky; Low in West	Morning Sky; Low in East
1999	late Feb. and early Mar.	mid- to late Aug.
2000	early to mid-Feb.	late July and early Aug.
	early to mid-June	mid- to late Nov.
2001	mid- to late Jan.	early to mid-July
	mid- to late May	mid- to late Sept.
2002	late Apr. and early May	mid- to late Oct.
2003	early to mid-Apr.	late Sept. and early Oct.

Note: Mercury is visible less than 20 degrees from the horizon; always observe from a location with an unobstructed horizon in the specified direction, 60 to 80 minutes after sunset or before sunrise. Data for northern hemisphere only.

	Evening Sky; Low in West	Morning Sky; Low in East
2004	late Mar. and early Apr.	early to mid-Sept.
2005	first half of Mar.	late Aug. and early Sept.
		first half of Dec.
2006	last half of Feb.	early to mid-Aug.
	mid- to late June	mid- to late Nov.
2007	late Jan. and early Feb.	
	late May and early June	early to mid-Nov.
2008	early to mid-Jan.	
	early to mid-May	late Oct. and early Nov.
2009	last half of Apr.	first half of Oct.
2010	late Mar. and early Apr.	last half of Sept.

Where to Find Mars

	Jan.	Feb.	Mar.	Apr.	May	June	July	Aug.	Sept.	Oct.	Nov.	Dec.
1998	Cap	—	—	—	—	—	—	—	Leo	Leo	Vir	Vir
1999	Vir	Vir	Lib	Lib	Vir	Vir	Vir	Lib	Sco	Sgr	Sgr	Cap
2000	Aqr	Psc	Psc	Ari	—	—	—	—	—	Leo	Vir	Vir
2001	Lib	Sco	Sco	Sgr	Sco	Sco	Sco	Sco	Sgr	Sgr	Cap	Aqr
2002	Psc	Psc	Ari	—	—	—	—	—	—	—	—	—
2003	Lib	Sco	Sgr	Sgr	Cap	Aqr	Aqr	Aqr	Aqr	Aqr	Aqr	Aqr
2004	Psc	Psc	Ari	Tau	Tau	—	—	—	—	—	—	—
2005	—	—	—	Cap	Aqr	Psc	Psc	Ari	Ari	Ari	Ari	Ari
2006	Ari	Tau	Tau	Gem	Gem	—	—	—	—	—	—	—
2007	—	—	—	Aqr	Psc	Psc	Ari	Tau	Tau	Gem	Gem	Gem
2008	Gem	Tau	Tau	Gem	Gem	Cnc	Cnc	Leo	—	—	—	—
2009	—	—	—	—	—	Ari	Tau	Tau	Gem	Gem	Cnc	Cnc
2010	Cnc	Cnc	Cnc	Cnc	Cnc	Leo	Leo	Vir	Vir	—	—	—

Note: Dashes indicate that the planet is in twilight glow and difficult to observe. Three-letter abbreviations are for the 12 zodiac constellations. Use all-sky charts in Chapter 4 to determine where Mars is seen in the sky for a specific month (i.e., the constellation through which it is passing).

CHAPTER EIGHT

MOON AND SUN

It is the very error of the moon;
She comes more near the Earth than she was wont,
And makes men mad.

William Shakespeare

Seventeenth-century Italian scientist Galileo was, so far as we know, the first to use a new invention called the telescope to peek at celestial objects. When he turned his instrument to the Moon in 1609, he was astounded at what he saw. "The Moon is not smooth and uniform," he wrote, "but is uneven, rough and full of cavities." With one glance, Galileo had smashed the centuries-old belief that heavenly bodies were perfectly formed and precisely spherical.

Times have changed, but at least part of the rush of amazement that Galileo must have experienced when he initially looked at the Moon through a telescope is felt by everyone who sees the lunar surface close up for the first time. Even with binoculars, the view is startlingly sharp and clear, with a few dozen craters and rugged mountain peaks plainly visible.

No matter what instrument is used, lunar detail is more distinct along the terminator—the line dividing the illuminated and the unilluminated portions of the Moon—because of the sharp relief effect caused by the shadows. Contrary to expectations, the full Moon is the worst phase to observe, because the strong relief effect is absent. As the Moon nears full, light-colored rayed craters stand out like splashes of white paint, but most other surface features are lost in the wash of light. The days around the first and last quarter are best for detailed viewing.

The Moon is so near that any telescope will show a wealth of detail. For example, a 50mm (2-inch) refractor at 50x will detect craters less than 10 kilometers across. With every increase in telescope size, up to about 10-inch aperture, more and more detail is revealed. The limit is about one kilometer, although shadows cast by features as small as 30 meters high can be glimpsed right at the terminator. Larger telescopes seldom reveal much more because of seeing limitations.

Astronomers have known for more than a century that the Moon's surface has been

Sunlight reflecting off Earth imparts a ghostly glow to the night side of the Moon. Known as Earthshine, the phenomenon is most prominent at crescent phase, as seen above. The bright object at upper left is the planet Venus.

an unchanging vista. The craters and plains we see are the same ones that Galileo looked at and our prehistoric forebears must have wondered about when pondering the nature of the silvery orb riding the night sky. But in that constant vista, there is much to see. A good telescope offers a lifetime of lunar exploration. There is a magical sense of discovery as you patrol the terminator while it slowly advances hour by hour and night to night.

The four main phases of the Moon—first quarter, full, last quarter and new—are enough detail for wall calendars, but backyard astronomers describe the phases in terms of the Moon's "age" in days after new phase (*new* is the point when the Moon is closest to being between Earth and the Sun). One complete lunar cycle from new to new takes 29½ days. First quarter occurs at around 7 days, full Moon at 14 or 15 days and last quarter at about 22 days.

There are two basic classifications of features on the lunar surface: the familiar craters and the darker plains. The craters are named after prominent philosophers and scientists of the past. The plains are officially called seas, or *mare* in Latin, because they were assumed to be bodies of water by Galileo and other 17th-century lunar observers. Mare Crisium, best seen when the Moon is three to five days old, is a particularly distinctive plain because of its dark tone and circular appearance. Current theory suggests that it was once a vast crater which flooded with lava early in the Moon's history.

Around first-quarter phase, age six to nine days, the Moon is at its prime for backyard skywatchers. About 35 percent of the surface consists of dark, smooth plains, many of which have high-flown names inherited from the 17th century. Mare Tranquillitatis, the Sea of Tranquillity, was the site of the first moonwalk ("Tranquillity Base here...the Eagle has landed"). The landing site is not far from the crater Maskelyne. Joined to Tranquillitatis is the smaller Mare Nectaris, whose southern edge overlaps the largely ruined crater Fracastorius. On the edge of Nectaris nearest the Moon's center are three huge craters that provide a perfect contrast to the smooth plains of Nectaris. This region is breathtaking when near the terminator.

The most impressive of the three craters is 100-kilometer-wide Theophilus, one of the truly great lunar craters. The walls of Theophilus soar to heights of more than 4,400 meters above its floor and cast magnificent shadows that intensify its dark, rugged appearance. When the Moon is five to six days old, the terminator plunges the crater

At full Moon, the white spoke-like rays emerging from the crater Tycho (bottom center) are obvious in binoculars.

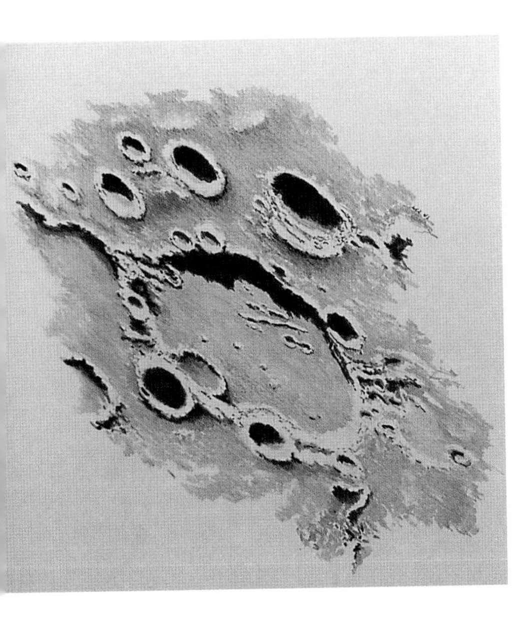

deep into shadow. I have caught it with only the towering central peak illuminated, the inky blackness giving an eerie and totally unearthly aura to the great crater. When Theophilus was formed a few billion years ago, it smashed down a wall of its nearest neighbor, the equally wide Cyrillus, whose ramparts are now only half as high. Forming the final member of the trio is Catharina, a crater about the same size as Theophilus but in worse shape, its crumbled walls only 3,000 meters higher than its rubble-strewn floor. These three craters, each in a different stage of decay due to the various forces that have modified the lunar surface throughout its history, provide a vivid cross section of crater forms.

Vast lunar plains like Tranquillitatis and Serenitatis are believed to be the filled-in remains of enormous craters that were blasted out of the lunar surface by impacting asteroids during the early days of the solar system. Rippling across these two Ohio-sized plains are ridges resembling waves of lava that cooled and then froze in place. This is almost certainly what they are. The low Sun angle makes the floor of Serenitatis look as if it is riddled with the underground trails of giant lunar rodents.

After first-quarter Moon (9 to 11 days old), the terminator sweeps over some spectacular lunar scenery. This phase, known as gibbous, which means humpbacked (from the convex curvature of the terminator), reveals the rugged southern section of the visible lunar disk. Although the craters are jammed together in bewildering profusion, the monster crater Clavius, 230 kilometers across, clearly stands out as the largest. Clavius, an easy target for binoculars, is so vast that it has two 50-kilometer-wide craters within it as well as a number of easily seen smaller ones. But these intruders barely make inroads into this colossal lunar feature. The giant walls of Clavius range up to five kilometers above its floor. Yet the slope of the walls is gentler than it appears. The walls are 50 kilometers wide at the base, so the wall-slope angle is only 10 degrees.

Before spacecraft mapped the Moon, sketching lunar features, above, was a prime activity among amateur astronomers. Today, it is a lost art, and the Moon is often passed over as a target of scrutiny. That's too bad, because the lunar surface is a wondrous alien landscape worthy of many hours of telescopic exploration. Right: This photograph of a section of the waxing crescent Moon displays Mare Serenitatis and Mare Tranquillitatis, two smooth lunar "seas."

Tycho, another prominent crater in this rugged sector, measures 85 kilometers across, with ramparts reaching more than four kilometers above the floor. The walls rise more steeply than those of most craters, giving Tycho a jagged appearance when the terminator is nearby. When examining the massive walls and sharp peaks of Tycho and similar craters, one gets the feeling that they must be spectacularly impressive from the surface of the Moon. Alien, yes, but compared with the Earth's major mountain ranges, like the Rockies, even the steepest crater walls are subtle hills. There may once have been such peaks on the Moon, but billions of years of impacting debris has hammered them down. In addition, the lunar surface material, which is comparable to lumpy garden dirt, tends to slump over time. This is evident in the terraced walls of craters. Tycho is believed to be the youngest of the large lunar craters, and consequently, its walls are among the steepest.

How young? Tycho was formed 109 million years ago by the impact of a five-kilometer-wide comet or asteroid. Just 44 million years later, an object twice that size clobbered Earth, with devastating consequences, almost certainly including the extinction of the dinosaurs. The accurate determination of Tycho's age comes from samples collected by the Apollo 15 astronauts, who explored a sector of the Moon 2,250 kilometers from the big crater. The samples that yielded the age were collected on a hill which was covered by material thrown across the lunar landscape from Tycho's explosive creation.

This material is visible in binoculars around full Moon, when Tycho appears to be at the hub of a system of white spokes resembling lines of longitude radiating from the pole of a globe. The impact that created Tycho hurled debris in all directions—in some cases, one-third of the way around the Moon. The rays emanating from Tycho are lighter than the older lunar surface they cover because solar ultraviolet radiation darkens Moon dirt as time passes.

Near the terminator when the Moon is 9 to 10 days old lies the splendid Copernicus crater, considered by many observers to be the most awesome lunar feature. At 93 kilometers in diameter, it is certainly not the largest crater on the Moon. What gives Copernicus prominence is its position in an otherwise flat lunar plain. The contrast is striking. The great walls of Copernicus and a blanket of debris splashed out around it make this the prototype of craters. Imagine the processes that occurred when a multibillion-ton asteroid slammed into the Moon, gouging out this feature perhaps 800 million years ago. When Copernicus is close to the terminator, ragged inky shadows from its peaks drape over the crater's interior and spill onto the surrounding plain.

Billions of years ago, the Moon was much closer to Earth than it is today. Consequently, our planet's gravitational pull on the Moon was stronger, raising substantial tides on the

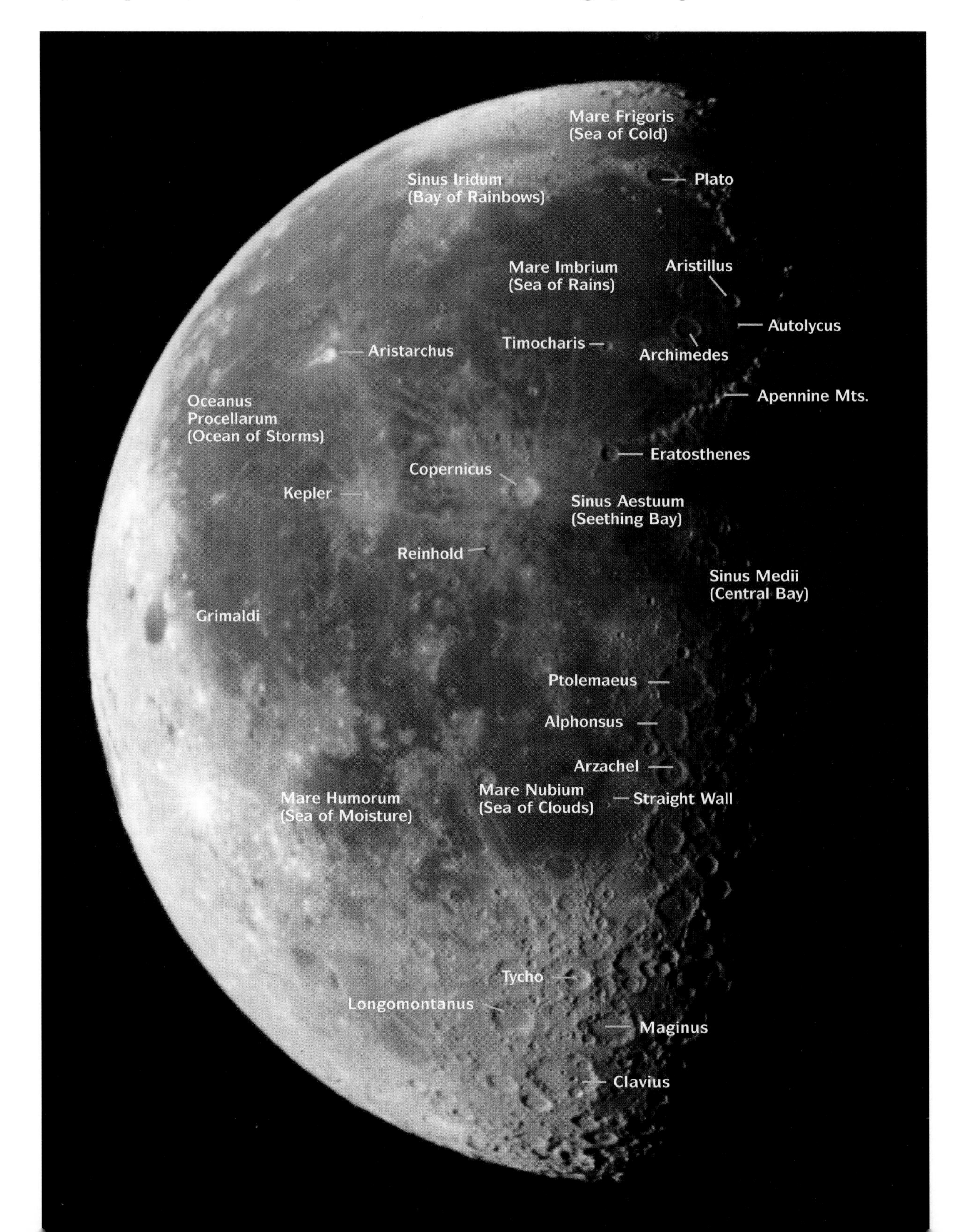

lunar surface. This force gradually retarded the Moon's spin on its axis, until the rate of spin exactly matched the Moon's orbital rate around Earth, leaving it with its slightly more massive side seemingly locked toward us. No matter what the Moon's phase or what season of the year, we never view the other side. Speculation about what secrets lay hidden on the other side of the Moon ended when the first spacecraft looped around our satellite in 1959 and found a cratered landscape similar to the side we see from Earth.

The hidden far side of the Moon is frequently—and incorrectly—called the dark side. The dark side is merely the portion of the Moon experiencing nighttime. The far side and the dark side coincide at full Moon. At other lunar phases, portions of the near side are also dark. The complete cycle of phases repeats every 29½ days, with all places on the Moon in darkness for half of that interval.

Probably the best-known but least-understood lunar phenomenon is the Harvest Moon—the full Moon that occurs nearest the autumnal equinox, the first day of autumn in the northern hemisphere, usually September 22. For northern-hemisphere observers, the full Moon nearest the autumnal equinox seems to linger in the sky night after night. To farmers of earlier generations, the Harvest Moon was an unexplained but welcome bonus of light.

From night to night, the Moon moves about 12 degrees eastward. As a result, the Moon rises in the east an average of 50 minutes later each day. However, the geometry of the orbit of the Moon related to the tilt of the Earth's axis results in the Moon moving in its orbit along a trajectory nearly parallel to our horizon on the days near the autumnal equinox. This

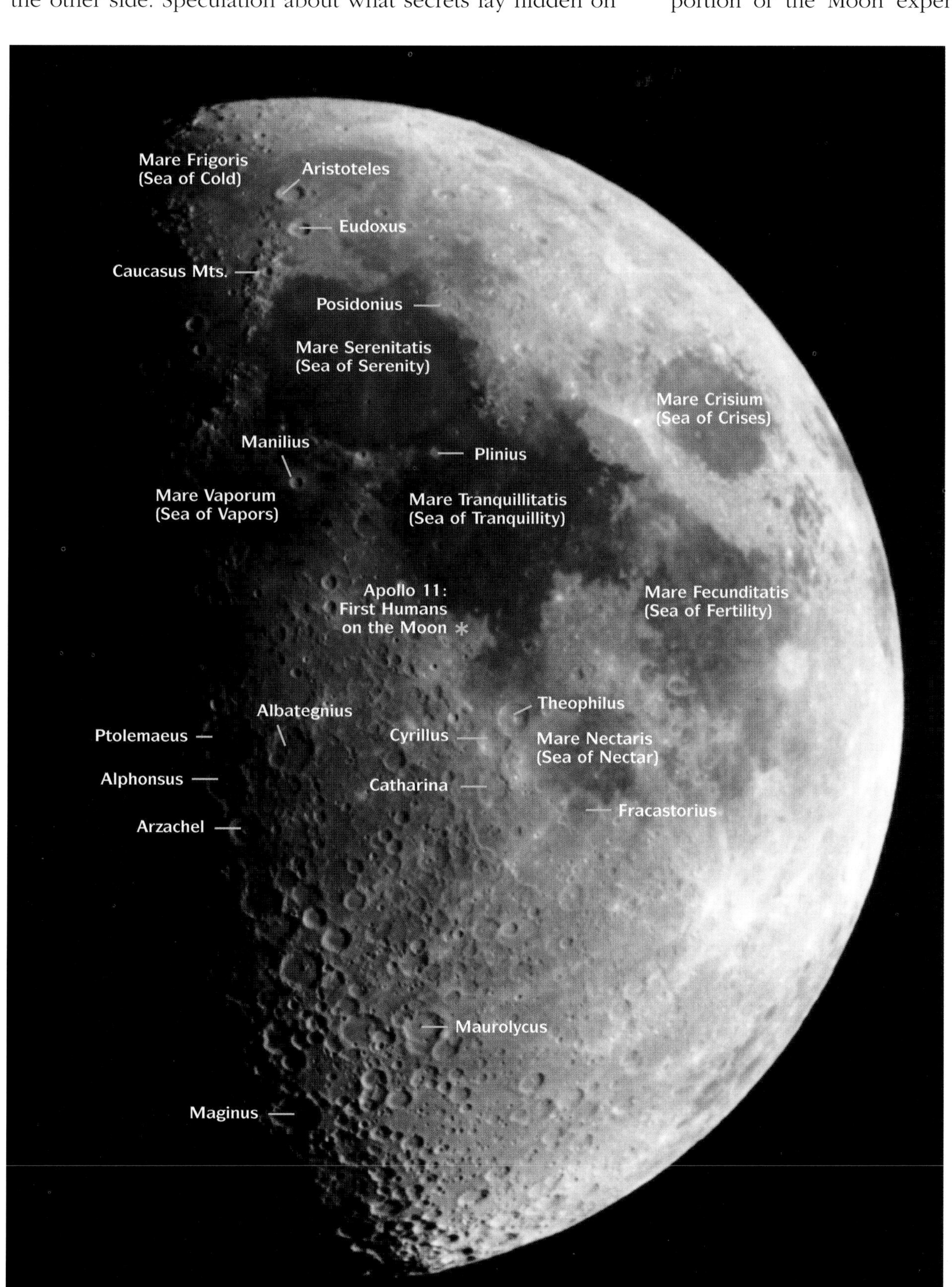

Binoculars, particularly when tripod-mounted, will show all the features identified in these photographs of the first-quarter Moon, left, and last-quarter Moon, facing page.

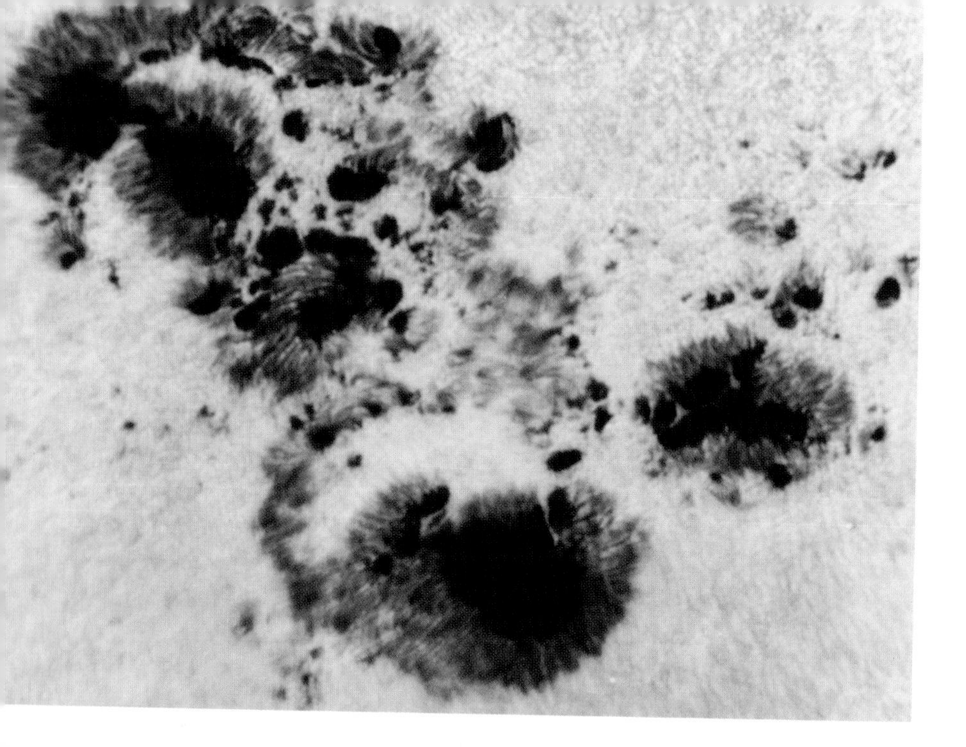

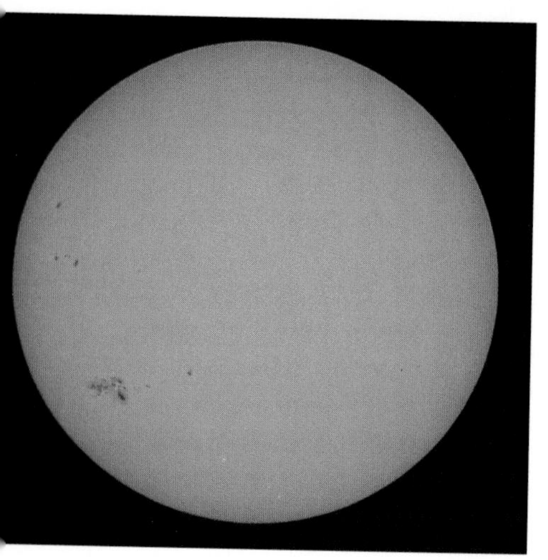

means that the time interval between successive risings is much shorter than average. In southern Canada and the northern United States, the Moon rises about 25 minutes later each night around the autumnal equinox. Thus for several nights near full Moon, there will be bright moonlight in the early evening around traditional harvesttime.

Sight-seeing on the Moon from your own backyard is one of amateur astronomy's most accessible pastimes. The Moon can be observed just as well from the city or the country. The late Isaac Asimov, a science and science fiction author who lived in a 33rd-floor apartment in Manhattan, viewed the Moon perfectly well with a telescope from his balcony. One early morning when he passed a window on his way to the kitchen, he had this reaction: "Looking out the westward window, I saw it: a fat, yellow disk in an even slate-blue background, hanging motionless over the city. I found myself marveling at Earth's good fortune in having a moon so large and so beautiful."

Observing the Sun

What would the average star look like close up? Probably not much different from the Sun, whose surface can be examined by any backyard astronomer. But this should not be attempted without taking strict precautions. Concentrated sunlight streaming through a telescope's ocular can cause blindness in less than a second.

There are two chief methods of solar observation that are perfectly safe. The first procedure requires a full-aperture solar filter, which intercepts the Sun's light before it enters the telescope, reducing the Sun's brightness by a factor of about 100,000 and thus bringing the light intensity down to a safe, comfortable level. Such filters are attached to the telescope in front of the objective lens or mirror.

Full-aperture filters are made of specially coated optical glass or highly reflective coated Mylar film. While more expensive, the glass filters yield close-to-true-color images of the Sun. The less costly Mylar filters usually give the Sun a bluish tint, although some close-to-true-color-transmission Mylar filters are available.

Full-aperture filters are usually not supplied as standard equipment with telescopes. However, some telescopes do come with a small filter designed to be placed at either end of the eyepiece—capped on the end closest to the eye or screwed or clipped onto the end that fits into the focusing sleeve. I do not recommend the use of these eyepiece filters. They quickly heat up from concentrated solar radiation near the focus of the telescope and, after a few minutes, crack or melt, allowing a sudden burst of intense radiation to strike the observer's eye.

The alternative to using filters is the indirect viewing method, whereby the solar image is projected by the telescope onto a white viewing card. This technique has an advantage over filtration because it does not require any accessory equipment. The telescope system is used as is. Binoculars stabilized on a tripod can also be used. The intensity of sunlight is more than sufficient to produce a readily visible image that can be sharply focused.

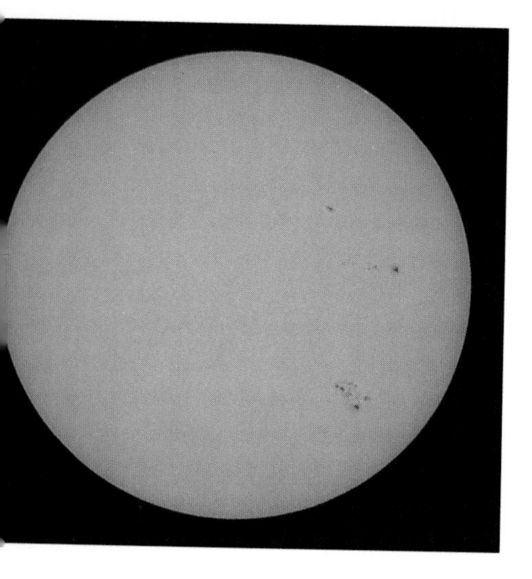

Properly filtered, a small telescope reveals impressive detail on the Sun, our nearest star. Sunspots, seen close up, top, are the most obvious activity. The full-disk photograph at left was taken the same day. The photograph in the middle, obtained a week earlier, shows the effect of the Sun's rotation. Less conspicuous detail, such as solar granulation (seen on the surface away from sunspots) and faculae (bright arcs near the solar limb), can be glimpsed during moments of good seeing.

Keep an eye on any children who are enjoying the activity. They are at the right height and have enough curiosity to try to look through the telescope. It is also a good idea to cover the objective of the finderscope, since it projects its own image. Projection usually offers less detail for solar-surface observation than the filtration method does, but it has the advantage of providing an image for several onlookers. Hold the projection card one to two feet behind the eyepiece to get the best combination of image brightness and size (a 3-inch diameter is good).

Small refractor telescopes are best for projection. Bigger instruments are badly affected by poor seeing caused by solar heating of both the instrument and the atmosphere. Telescopes larger than 70mm aperture *must* have the aperture restricted to 60mm or less for solar projection. Cut a circular hole in a piece of cardboard, and tape it in front of the instrument. This reduces the seeing effects and prevents the optical system from overheating, which can be especially dangerous with Schmidt-Cassegrains. Intense heat can also damage expensive eyepieces, another good reason to reduce the aperture.

Don't try to aim at the Sun by looking along the telescope tube! Look at the shadow. Adjust the aim point until the tube makes the smallest possible shadow on the ground. The Sun should then be nearly centered in the telescope, and probably only a slight adjustment will then be needed to bring the image onto the projection screen.

For unaided-eye observation, a conveniently available and perfectly safe filter is the glass insert from welders' goggles. Ask for a No. 14 welders' plate at a local welding-supply outlet. It comes in a handy 2-by-4-inch rectangle that permits direct viewing of the Sun with both eyes. This filter material can be taped over the objective lenses of binoculars. However, it is *not* safe to use at the eyepiece end, where the light is concentrated. To be effective, *all* solar filters must diminish the solar intensity *before* it enters the optical system.

With the unaided eye protected by a welders' filter, tiny black spots can often be picked out on the dazzling solar disk. These are sunspots. The fact that the larger spots can be seen with the unaided eye and a proper filter is one of the best-kept secrets of backyard astronomy. But anyone with average vision should be able to see a major spot if it is there.

Sunspots have been observed with the unaided eye for at least a thousand years. Chinese astronomers noted black speckles on the solar disk near sunset, when its image is reddened and subdued by atmospheric absorption. Do not try to repeat the Chinese observations if you have to squint. The Sun must be *deep* orange.

Properly filtered binoculars almost always reveal a sunspot or two, and a small telescope will show all the spots on the solar face—perhaps a dozen or more. A magnification of 40 or so is all that is needed for a good view of the entire Sun. Higher powers show less than the full solar disk, and when the projection method is used, they render the Sun's image too dim to be viewed clearly on the projection screen. It is amazing what can be seen with low magnification. Since the Sun is a colossal globe of gases (mostly hydrogen), the visible outer surface is constantly changing.

A sharply focused telescopic image of the solar disk will

Screw-in eyepiece solar filters, top left, sometimes supplied with beginners' telescopes, should NOT be used. They can crack and allow damaging amounts of concentrated sunlight to hit the eye. Projection of the Sun's disk, top right, or full-aperture solar filters are the recommended techniques. Special narrow-band hydrogen alpha filters, although relatively expensive, disclose far more detail on the Sun's surface, left, than do normal solar filters.

reveal several sunspots, some bright patches and a gradual darkening toward the edge, called limb darkening. With good seeing conditions, the Sun's surface appears mottled or granulated, like leather when closely examined. This solar granulation is real and consists of cells of rising gas boiling like water in a pot. About the size of Lake Superior, a solar granule changes its shape in a matter of minutes. The specific change is not evident, because it is impossible to concentrate on a single granule for more than a few seconds. The pattern overwhelms each individual spot. The darkened limb of the Sun is usually flecked with irregular, bright splotches, called faculae. These are clouds of hydrogen, often associated with sunspots, surging above the Sun's surface. The brighter solar surface renders them invisible near the center of the disk.

The major fascination of the Sun lies in those black blemishes, the sunspots. The spots themselves are actually cooler areas on our star, about 1,500 Celsius degrees cooler than the 5,500-degree-C solar surface. Being cooler, the spots are darker, but they appear black only by contrast. They are actually a light brown color and would appear so if seen alone.

There are two parts to a sunspot: a black, virtually featureless interior, called the umbra, and a grayish, feathery-structured zone around the umbra, called the penumbra. Combined, these two regions come under the general designation of a sunspot. Sunspots can be 10 or more times the size of Earth. An average sunspot is at least as big as our planet.

Like all visible solar features, the spots are caused by intense magnetic fields that coil within the Sun and break out through the surface in largely unpredictable ways. A sunspot is the focus of a magnetic-field breakout, a region where the flow of energy from within the Sun is restricted, hence the less luminous appearance. An individual spot emerges as if from nowhere over a period of a day or two. Spots can last for days or weeks. Some of the largest sunspots remain visible for several months.

The number of spots varies from year to year over a fairly well-defined 11-year period known as the sunspot cycle. These cycles do not exactly repeat one another; some are more "spotty" than others. In general, the rise to maximum is more rapid (about four years) than the drop to minimum (about seven years). The most recent minimum occurred in 1997. The mechanism behind the sunspot cycle remains a scientific enigma. Spots do not appear over the Sun at random. They are usually confined to a region 15 to 40 degrees north and south of the solar equator. At sunspot maximum, the spots are generally closer to the equator than they are at other times in the cycle.

Following the progress of sunspots over a period of weeks or months is easy. Each time the Sun is visible, simply place an observing sheet on a clipboard, hold it behind the telescope and make a sketch of the sunspots' positions. This should result in an accurate enough drawing to show positional changes, as well as the growth and decay of individual spots. From one clear day to the next, these drawings will track the spots as the Sun's rotation carries them around the visible disk. (The Sun rotates once in about 27 days.) To ensure that each picture has the same orientation, note the direction of the Sun's drift through the *center* of the field of view caused by the Earth's rotation. The point on the Sun's disk that first drifts out of the field of view will establish a reference mark on the circumference of each circle, permitting the proper relationship of one drawing to the next.

Solar observing ranks as one of the most pleasant astronomical activities, because it is conducted on sunny days and requires only modest equipment. The joys of solar observing have really hooked some amateur astronomers. Not content with standard methods of watching the Sun, they equip their telescopes with narrow-band filters that permit the ultimate—watching solar prominences, those "flames" seen in the most spectacular photographs of the Sun. These filters range from several hundred to several thousand dollars. The performance of the more expensive ultra-narrow-band filters is awesome. The first time I saw the Sun through one, I could not tear myself away. Our familiar star was transformed into a ball of roiling plumage as intricate in its detail as is the rugged face of the Moon.

Full-aperture solar filters are available for both binoculars and telescopes. They come in two basic types: aluminum-coated Mylar, top, and metal-coated glass, right. Glass solar filters provide a pleasing yellow hue to the solar disk.

The Moon Illusion

Have you ever noticed how the Moon appears larger when it is near the horizon than when it is overhead? The difference is so apparent, it seems almost impossible that it is not real. Yet how could it be? The Moon is no closer to us on the horizon than it is overhead. Actually, it is about 6,500 kilometers farther away, as we must look across the radius of Earth.

The same effect occurs with the Sun. It seems enormous as it dips below the horizon, a fiery ball reddened by intervening dust and particles in the Earth's atmosphere. At these times, the Sun is distorted to an oval by the refractive properties of the atmosphere which bend light rays, similar to the way that a straight stick looks bent when partially submerged in water.

Could this refraction be enlarging the Sun's apparent size? Simple tests have actually shown that the reverse occurs. The atmosphere acts like a weak lens, compressing the Sun's vertical dimension and giving it an oval shape that is smaller than the circular disk would appear. The same phenomenon can be seen when the full Moon is close to the horizon. It looks like a huge cosmic pumpkin, the reddening caused by the same atmospheric dust and haze that redden the Sun.

Since neither atmospheric refraction nor changes in distance make the Moon or Sun bigger on the horizon than when either is higher in the sky, why do they seem so large? The effect is particularly noticeable with the Moon because it can be compared with its appearance higher in the sky, whereas the Sun is too bright for easy comparison. When people unfamiliar with astronomy are quizzed about the Moon's size, virtually all of them insist that our satellite is bigger when it is close to the horizon. Why?

The Moon illusion, or horizon illusion as it is sometimes called, was recognized as an enigma as long ago as 350 B.C., when Aristotle incorrectly attributed it to atmospheric "vapors" that distort images close to the horizon. Around the year 1000, Arabian physicist Ibn Alhazan offered the first modern explanation. He suggested that a familiar background, such as distant trees or houses, provides a frame of reference not available when the Moon is overhead. Since the Moon looks huge by comparison with these familiar objects, the mind insists that it is vast in size.

Alhazan's theory sounds plausible, but it fails to explain why the same effect occurs with a perfectly flat desert or ocean horizon. The illusion works even in a planetarium. The projected image of the Moon seems bigger near the horizon than when it is higher up on the planetarium dome, although the lunar image may then be closer to the observer. Obviously, another mental factor is coming into play in addition to comparisons with objects on or near the horizon. It was uncovered in 1959 during a test at the University of Wisconsin.

In this test, a 20-inch disk was suspended overhead 85 feet above the ground, and another disk the same size was placed 85 feet away on the horizon. People of all ages were asked to stand at a point equidistant from the two disks. The observers did not know that the disks were identical and therefore should appear exactly the same. Everyone tested thought that the disk on the horizon was the larger of the two. Young children most overestimated the difference, some saying that the disk on the horizon looked three or four times bigger than the one overhead.

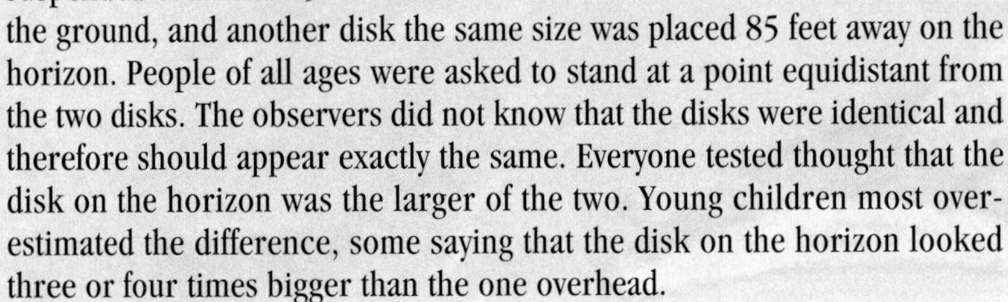

Somehow, looking up has something to do with the illusion. As a further test, other researchers put volunteers in a dark room with a disk straight ahead and an identical disk at the same distance overhead. Again, everyone thought that the overhead disk was smaller. So apparently, two factors are involved: (1) association with the distant horizon; and (2) looking straight ahead as opposed to looking nearly overhead.

Yet there is more to it than that, but nobody is sure just what it is. Even though I am fully aware of the Moon illusion and its various explanations, I still see the full-blown effect. It's one of the most powerful illusions in nature.

But there are ways of diminishing the illusion. When the Moon is near the horizon, try looking at it through a tube. Without the horizon reference, it seems smaller. Another method of countering the effect is to lie down and look at the Moon near the horizon from a position flat on the ground. The Moon does not appear nearly as large as it does from a standing position, particularly if your neck is craned to look at it over your head or down toward your feet. Or try standing and bending over from the waist to look at the Moon from between your legs. Again, it appears much smaller.

If all this still sounds unconvincing, here is the final test: An aspirin tablet held at arm's length is only slightly larger than the Moon. It will cover it nicely, whether the Moon is hovering over the horizon or riding high in the night sky. Try it.

CHAPTER NINE

SOLAR AND LUNAR ECLIPSES

It is only during an eclipse that the man in the moon has a place in the sun.

Anonymous

I saw my first total eclipse of the Sun in February 1979 as a member of The Royal Astronomical Society of Canada's eclipse charter flight, which flew from Toronto to Gimli, Manitoba, the morning of the eclipse. We landed as planned on an unused air-force-base runway two hours before the Moon took the first notch out of the Sun. Although clouds and snow were predicted, the sky was clear, and the jubilant troop of eclipse buffs unloaded about 75 telescopes and more than 100 cameras and set up on the tarmac.

As the eclipse progressed, the scene began to resemble the climax of the movie *Close Encounters of the Third Kind*, which also involves a battery of equipment on a runway, with scientists and others awaiting the approach of extraterrestrial spaceships. Like the scientists and technicians in the movie, we were not disappointed. As one eclipse watcher said later, it was as if a god had decided to make an appearance for two minutes, and we knew he was coming.

I was completely unprepared for the overwhelming power of the eclipse. About two minutes before totality, the Sun's image was reduced to a thin slice along the rim of the black disk of the Moon. I knew that the Sun would be gone in a few seconds and that we would be plunged into darkness. Then, like a vast, diffuse storm cloud, the Moon's shadow suddenly appeared in the west, growing larger by the second.

With surprising suddenness, the shadow of the Moon swept over us, the last rays of sunlight disappeared, and the Sun was instantly transformed into an awesome celestial blossom—the black disk of the Moon surrounded by streamers of the Sun's atmosphere, the corona.

Peeking around the black disk and plainly visible to the unaided eye were a half-dozen solar prominences, like fingers of frozen fire. These surges of hot hydrogen, flamelike in appearance, constantly lurch from the surface of the Sun, propelled by immense magnetic fields. I knew they were

The Sun's pearly corona, our star's tenuous atmosphere, is seen in its full glory during the total solar eclipse of February 26, 1998. Facing page: An annular eclipse occurs when the Moon is in the part of its orbit that carries it too far from Earth to cover the Sun completely.

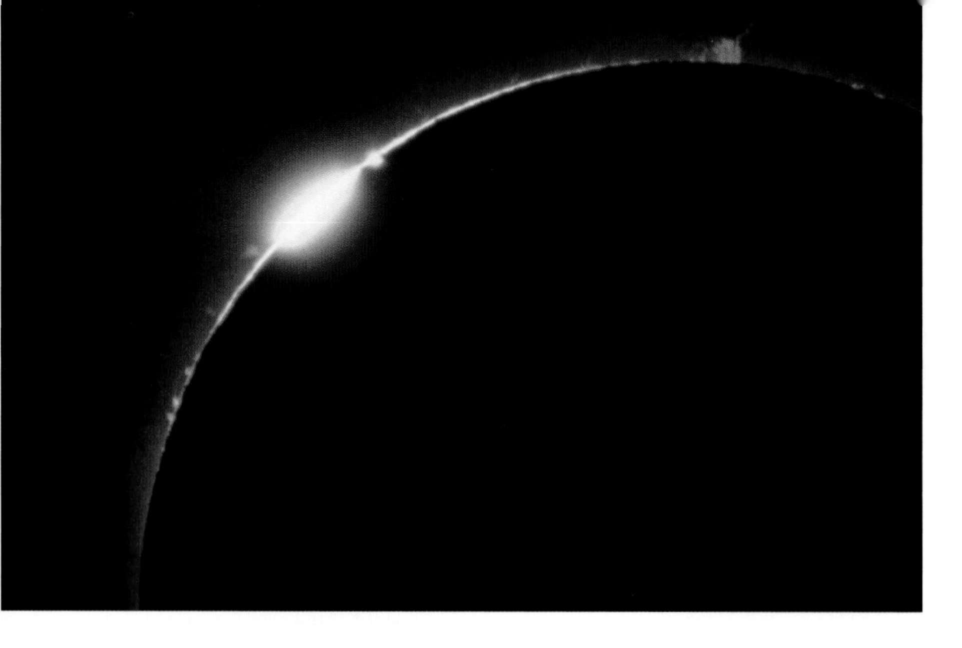

there, but I did not expect to see them so plainly. (Neither did anyone else, for it turned out that the prominences have never been more spectacular than they were during that eclipse.)

I was so awestruck by all of this that I became incapable of performing even the most rudimentary tasks. I realized immediately that I would never be able to photograph the event—I wanted every second of viewing time to enjoy the image. I wrenched the camera off my telescope and quickly slipped an eyepiece into the focuser. Then I gazed upon what has to be the most stunning astronomical spectacle that I have seen in more than four decades of skywatching.

Through the telescope, I could see detailed structure within the prominences, including one that was completely detached from the surface like a suspended ball of fire. Judging from its apparent size, it must have been several times wider than the entire Earth. The most striking feature was the vast range of delicate hues and intricate detail in the corona, the beautiful halo that surrounds the eclipsed Sun. It ranged from a gorgeous pinkish orange close to the Sun to various shades of pale yellow, pink and blue farther from the disk. Its overall brightness was comparable to that of the full Moon, an eerie yet mesmerizing cosmic flower. The Sun's magnetic field twists the corona into feathering arches and swirls that I was able to follow almost a full Sun's diameter from the surface of our star.

The Moon's black disk looked like a hole in the sky, with a ghostly aura around it and little tongues of pink flame licking its black circumference. Because of some pale ice-crystal cirrus clouds, the sky did not grow as dark as it had during some previous eclipses, or so I was told by several of my eclipse-chasing colleagues. Nonetheless, the darkness during totality and the color accompanying it were like a bizarre twilight, with simultaneous dawn and dusk—light in every direction around the horizon and darkness higher in the sky.

In the last few seconds of totality, a glimmer of Sun seared between mountain ranges at the edge of the Moon's disk. This is the aptly named diamond-ring effect. The diamond ring lasted for several seconds, growing from a starlike spot to a dazzling glow. Then Baily's beads appeared as more Sun leaked through the ridges on the Moon's limb. After that, the Sun was too bright to look at, and it was back to the welders' filters. But by this time, everyone in our group was shouting, cheering and applauding. I was speechless, beside myself with wonder at such an extraordinary visual symphony.

That 1979 eclipse was the last time the Moon's shadow will touch Canada or the United States until 2017 (marginal exception: the Arctic eclipse of August 1, 2008). On average, a particular place on Earth is treated to a total solar eclipse only once in 360 years. Traveling to where the eclipse is happening rather than waiting for it to come to you is the only reasonable strategy. I have journeyed to view totality four times, although only two occurred in cloudless skies—the 1979 event and one in the Caribbean in February 1998. Some eclipse buffs have racked up 10 or more successful expeditions. About 30,000 enthusiasts traveled to the Caribbean in 1998, most of them watching from the decks of cruise ships that positioned themselves like a line of floating hotels along the path of totality.

The arrangement necessary for a total solar eclipse is an exact alignment of Earth, Moon and Sun, with the Moon casting its shadow on Earth. This is a fairly frequent event that occurs almost every year. However, the problem for those who wish to view a total eclipse of the Sun is the small size of the Moon's shadow. By the time the Moon's shadow reaches Earth, it is usually less than 200 kilometers wide. One must be within this shadow in order to see the "darkness at midday," which lasts from a few seconds to seven minutes.

Only from within the narrow shadow is the Sun entirely blotted out by the Moon. The Sun is replaced by a black disk—the nightside of the Moon. The exquisite beauty of a total eclipse of the Sun can be seen quite safely with the unaided eye or a telescope. (To make this point perfectly clear, no eye protection is needed during the few minutes of totality, when the disk of the Sun is completely obscured. However, the filter methods described in Chapter 8 are always necessary during any other aspect of the eclipse.)

The key factor is that you must be within the Moon's shadow. Close doesn't count. If just a little bit of the Sun remains peeking around the Moon's disk, none of the spectacular corona is seen. Because they are visible from a much wider zone, partial eclipses are more common. But a partial eclipse of the

The diamond-ring effect, seen just before the Sun is completely covered by the Moon (and again at the end of totality), lasts just seconds, as tiny portions of the Sun's disk peek through mountains and craters at the edge of the Moon.

Flamelike solar prominences, the largest of which rival the size of Earth, appear pinkish to the eye, adding to the spectacle. This photograph of the February 26, 1998, solar eclipse was taken unfiltered through a 5-inch apochromatic refractor.

Sun in no way compares with a total eclipse. During most partials, the dimming of the sunlight is hardly noticeable. Observed through proper filtration, the Sun appears to have a nibble taken out of it, like a bite from a cookie. If an image of the Sun is projected telescopically or if appropriate filtration is used, the event is interesting and worth observing (see table on page 152).

Why do eclipses of the Sun not occur every month at new Moon, when the Moon is between Earth and the Sun? Most of the time, the Moon passes just above or below the Sun and does not obscure it from any position on Earth. The reason for this is that the Moon's orbit is slightly inclined to the ecliptic (by five degrees), which means that except when the apparent paths of the Sun and the Moon intersect, the Moon is either below or above the Sun, so no eclipse occurs. The same geometry applies to a lunar eclipse.

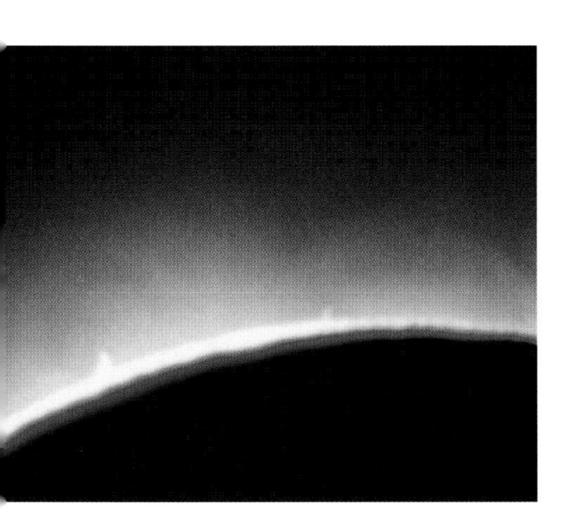

The earliest recorded solar eclipse probably occurred on October 22, 2137 B.C. The ancient Chinese chronicle *Shu Ching* tells of two royal astronomers, Hsi and Ho, who were so "drunk in excess of wine" that they failed to warn the populace of the impending darkness. The unexpected eclipse so frightened the people that they stampeded through the streets beating drums to frighten away the dragon devouring the Sun.

According to ancient Chinese law, any error by an astronomer in predicting eclipses would bring the gravest consequence. If the astronomer's forecasts were "behind the time, he would be hanged without respite." Apparently, that is what happened to Hsi and Ho for drinking on the job—or so the legend goes—but the tale also suggests that the Chinese could predict eclipses more than 2,000 years before the same method was practiced by the Greeks.

It used to be only scientists who journeyed around the globe to gaze at the midday night. The eclipse over Manitoba on June 18, 1860, is an example of the incredible effort that went into such academic jaunts. Simon Newcomb, later to become the most prominent astronomer of his time, headed an expedition from Boston to The Pas, in northern Manitoba. The trip took five weeks by steamer, covered wagon and canoe. Violent rainstorms and generally bad weather delayed the latter parts of the journey, which could be made only by canoe and portage.

Fearing that they would not make it to the selected site on time, Newcomb persuaded the hired voyageurs to paddle for 36 hours straight to get within the belt of totality. The effort was in vain, for Newcomb and his two assistants simply sat at their telescopes looking at the clouds. And as if nature conspired

Above left: Seen during a total solar eclipse as graceful loops and arcs, intense magnetic fields reach up from the Sun's surface, adding structure to the corona. Above: From southern California on January 4, 1992, the setting Sun was gradually disappearing as the Moon crept in front of it. At sunset, the event climaxed with a striking annular eclipse, as shown in the photograph on page 146.

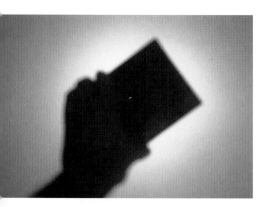

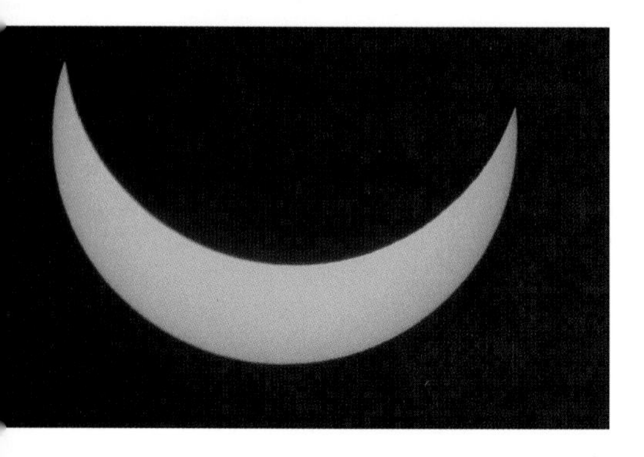

to add a final insult to the frustrated group, the clouds parted just a few minutes after the eclipse had ended.

Eclipses were of enormous scientific importance at one time. Observations that could not be made at any other time were possible during the reduction in solar illumination. Einstein's theory of general relativity, which predicted that light would be deflected in a strong gravitational field, was confirmed in 1919 by observing the position of stars near the eclipsed Sun. Stars normally invisible due to the Sun's brightness were seen briefly during the period of totality. Photographs of their positions revealed that their light was shifted by precisely the amount predicted in Einstein's theory.

So far as we know, the total-solar-eclipse phenomenon is unique to Earth. It happens because the Sun is about 400 times the diameter of the Moon but 400 times farther away. Thus they both appear almost exactly the same size. Nowhere else in the solar system does this specific arrangement occur. The satellites of Mars, as seen from the surface of that planet, are too small to cover the solar disk. None of the satellites of Jupiter and Saturn are the same apparent size as the Sun, being either significantly larger or smaller. At any rate, those planets are 5 and 10 times farther from the Sun, respectively, and any eclipse effect would be much diminished.

Coincidence is truly the only explanation, because the Moon has been gradually increasing its distance from Earth over the last three or four billion years. In about 200 million years, total eclipses of the Sun will no longer occur, because the Moon will be too far away from Earth. All solar eclipses will then be annular, that is, a small amount of Sun will peek around the edge of the lunar disk, ruining the awesome appearance of the event as currently witnessed. Today, Earth and the Moon are at the perfect balance, and we are enjoying the maximum effect of an incredible cosmic coincidence.

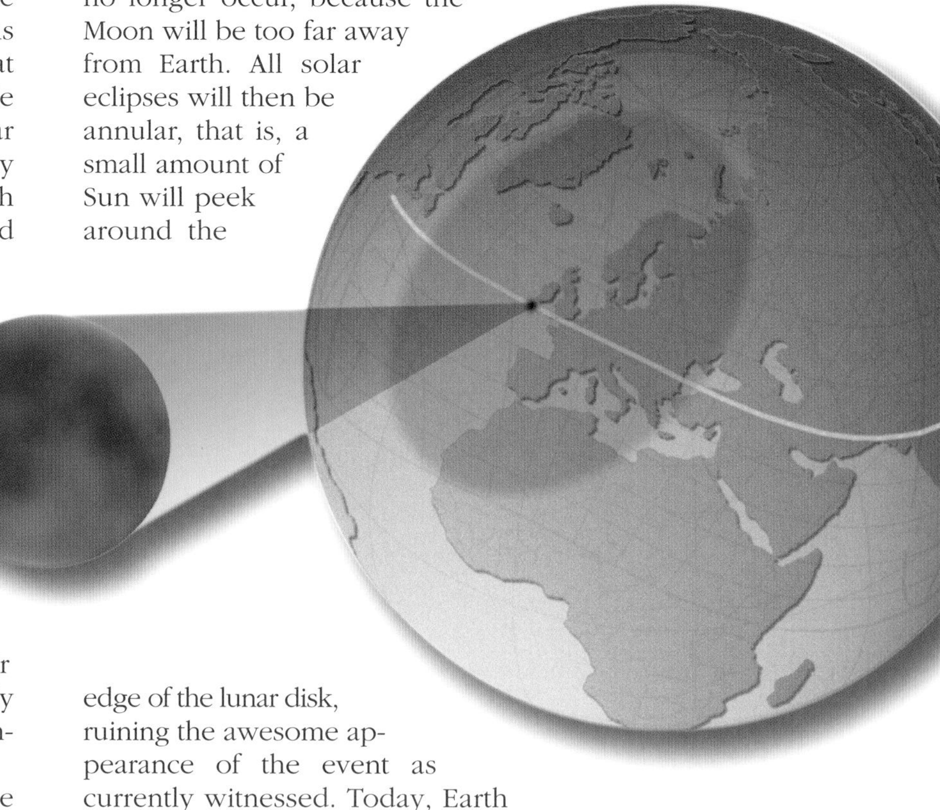

Top left: A durable and completely safe filter for naked-eye viewing of the partial phases of a solar eclipse is the #14 welders' filter plate, available at welding-supply shops. The filter gives the Sun a light green cast, center left. Telescope solar filters, such as the ones shown in Chapter 8, offer views of the partial phases, like the one at bottom left. Right: The path of totality of the August 11, 1999, total solar eclipse sweeps across heavily populated parts of Europe.

Lunar Eclipses

In 1503, Christopher Columbus and his crew found themselves stranded on the island of Jamaica, their ships damaged beyond repair. The native Arawak had decided that they were not going to supply any more food in exchange for baubles and trinkets. While the weeks dragged on, morale plunged as the crew was forced to live off the land.

Scanning his navigational tables, Columbus noticed that a total eclipse of the Moon would occur on February 29, 1504. He craftily devised a plan. The night of the lunar eclipse, he announced to the Arawak that the Almighty was frowning on the treatment they had been giving him and his crew. Gesturing to the sky, he reported that God had decided to remove the Moon as a sign of his displeasure. Within minutes, the Earth's shadow began to steal across the Moon's face. According to Columbus's diary, the theatrics worked. The Arawak promised to provide all the food the sailors would need if God would only bring back

The Eclipse Cult

Their numbers ebb and swell from one solar eclipse to the next, but they are always there, enduring any hardship or expense to spend a few hundred seconds standing under the Moon's shadow. They are the eclipse chasers. No place is too remote: southern Australia in 1976, India in 1980 and Java in 1983. With almost religious fervor, they converge on whatever location the celestial geometry dictates in order to witness nature's superspectacular.

Sometimes, their efforts go unrewarded. Whenever plans for an eclipse expedition are being made, amateur astronomers are reminded of the late J.W. Campbell, an astronomer at the University of Alberta. He traveled around the world to see 12 total eclipses during the first half of this century and was clouded out every time.

The success rate in recent years has been much better, because modern weather records allow more judicious site selections and because today's transportation methods offer more flexibility for the expedition. But modern eclipse chasers have their own misadventures.

During his globe-trotting to see 14 total eclipses (only one was completely clouded out), Vancouver astronomy enthusiast Ian McLennan has had his share of unusual mishaps. At a remote site in Kenya the night before an eclipse, a ferocious windstorm whipped the area where his group had camped. After nightfall, McLennan stepped outside his tent for a moment. When he returned, the tent had blown away. "I never did find it," he says. But he did see the eclipse.

Less fortunate was a planetarium director who, like thousands of others in June 1973, had booked a suite on one of several cruise ships specifically chartered to be under the Moon's shadow in the Atlantic Ocean at the critical time. Our hapless eclipse buff had to go below to use the men's room a few minutes before totality. But somehow, he got swallowed up in the labyrinth of hallways in the bowels of the ship. By the time he was able to find his way back on deck, the total eclipse was history.

McLennan's eclipse travels took him to eastern Quebec in 1972, where he had a revealing insight into human nature. People had lugged equipment from all over the globe to observe the eclipse, but when totality arrived, the clouds were too thick to see anything. "We did get plunged into blackness, of course," recalls McLennan, "and we noticed the weird calm that always accompanies totality. Yet on the Trans-Canada Highway nearby, the truckers simply turned on their headlights and kept going! Very few of them stopped to see what was causing the darkness at midday. And we had come from all over the world to experience it."

Eclipse addiction is not limited to amateur astronomers; it afflicts a wide cross section of nature lovers. As one veteran eclipse chaser put it: "It is highly contagious and requires only a single exposure to an unobstructed total eclipse."

Eclipse chasers spare no effort to travel to the remotest parts of the globe to stand in the Moon's shadow. They readily admit being addicted to the magnificence of a total eclipse of the Sun. The photographs on this page were taken on the Caribbean island of Curaçao during the eclipse of February 26, 1998. More than 30,000 people traveled to the Caribbean specifically to witness the phenomenon.

the Moon. These histrionics probably saved Columbus and his men from starving to death before they were finally rescued and returned to Europe.

An eclipse of the Moon can occur only at full Moon and only when Earth is exactly between the Sun and the Moon. When these conditions are met—usually twice a year—the Moon is engulfed in the Earth's shadow for up to 1½ hours. Unlike a total solar eclipse, which is visible only from a restricted zone of totality, a lunar eclipse is seen from the entire nightside of Earth, providing millions of front-row seats for interested observers.

Our satellite is never completely blacked out during a lunar eclipse. Sunlight diffusing through the Earth's atmosphere bathes the Moon in a dull glow that reduces it to about one ten-thousandth the normal brightness of the full Moon. The same principle causes the early-evening sky to remain relatively bright, even though the Sun is below the horizon.

The shadow's darkness varies with the amount of cloud, dust and pollution that is suspended in the Earth's atmosphere at the time of the eclipse. Occasionally, the obscuration is sufficiently dense

Future Total Solar Eclipses

Paths of totality are typically 100 to 200 kilometers (60-120 mi) wide; for maps and other details of each eclipse, consult astronomy magazines up to 18 months before eclipse day.

1999, August 11: totality runs through southwest England, northern France, Germany, Austria, Hungary, Romania and Turkey (path shown in illustration on page 150)
2001, June 21: South Atlantic and south-central Africa
2002, December 4: Mozambique and South Pacific
2005, April 8: South Pacific
2006, March 29: north-central Africa and Turkey
2008, August 1: Canadian high Arctic and Siberia
2009, July 22: China and equatorial Pacific

The next easily accessible total solar eclipses over Canada or the United States are a long way off—August 21, 2017, and April 8, 2024—but in both cases, the Moon's shadow will sweep from one side of the continent to the other.

Future Partial Solar Eclipses

Visible From North America

Times when partial eclipses occur can vary by several hours, depending on the observer's location; check almanacs or astronomy magazines for details.

2000, December 25: northeastern North America (60% in Great Lakes area around 2:35 p.m., EST)
2001, December 14: western North America
2002, June 10: western North America
2005, April 8: south of a line from San Diego to Philadelphia

Future Lunar Eclipses

Visible From North America

Times given are for mideclipse; begin watching at least one hour earlier to see the complete event; convert to your time zone, if necessary.

1999, July 28: partial; visible only from western third of continent (4:33 a.m., PDT)
2000, January 20: total; visible over entire continent; along with total eclipse of 2004, this is the best of the decade (11:43 p.m., EST)
2000, July 16: total only for Hawaii; seen as partial from West Coast (6:55 a.m., PDT)
2001, January 9: end of totality visible from Atlantic Canada; partial at dusk from New England, Quebec and eastern Ontario (3:20 p.m., EST)

2003, May 15: total; visible over entire continent except extreme northwest (11:40 p.m., EDT)
2003, November 8: total; visible over entire continent except West Coast (8:19 p.m., EST)
2004, October 27: total; visible over entire continent (11:04 p.m., EDT)
2007, March 3: total for Great Lakes and eastward; partial for rest of continent except extreme northwest (6:20 p.m., EST)
2007, August 28: total for Rocky Mountain region and westward; partial for rest of continent except extreme northeast (3:37 a.m., PDT)
2008, February 20: total; visible over entire continent (10:26 p.m., EST)
2010, December 21: total; visible over entire continent (3:17 a.m., EST)

to make the Moon disappear, while at other times, the shadow imparts only a pale rusty hue to the Moon. Eruptions from the Mount Pinatubo volcano in the Philippines in June 1991 spewed vast quantities of dust and sulfuric-acid haze into the upper atmosphere that spread throughout the northern hemisphere during the following months. The Pinatubo particles absorbed sunlight, cooling much of the northern hemisphere by a few degrees and producing a totally black shadow for the partial lunar eclipse of June 1992 and a dull gray shadow during the total lunar eclipse six months later. The Moon was so dim during that eclipse, I estimated that its overall light at mideclipse was equivalent to a fourth-magnitude star.

The Moon marches eastward in the sky a distance equal to its own diameter in about an hour as it moves in its orbit around Earth. The Earth's shadow is just under two Moon diameters wide, so if the Moon passes through centrally, it can be totally within the shadow for nearly two hours. Sometimes, our satellite just skims the shadow, and a partial eclipse results. The contrast between the bright Moon and the dark shadow during a partial eclipse makes the actual tone of the shadow difficult to determine, and the event is far less impressive than a total eclipse.

A total lunar eclipse unfolds as follows: About 20 minutes before the Moon is scheduled to enter the shadow zone, the eastern edge of the Moon becomes slightly dusky, indicating that the shadow region is nearby. However, when the edge of the Moon actually contacts the shadow, the darkening effect is unmistakable. During a total eclipse, the Moon takes about an hour to slip into the shadow. Once it is fully immersed, the total eclipse begins and can last anywhere from a few minutes to 1½ hours. The Moon swings out of the Earth's shadow during the final hour of the event.

During the 10 minutes centered on midtotality, try a visual estimate of the darkness of the eclipse using a scale developed by French astronomer André-Louis Danjon. This scale is intended for unaided-eye estimates:

Zero on Danjon's scale is a very dark eclipse; the Moon is practically invisible, especially during midtotality.

One is a dark gray or brownish eclipse; the lunar details are distinguishable only with difficulty.

Two is a deep red or rust-colored eclipse, with the central part of the shadow very dark and the outer edge relatively bright.

Three is a brick-red eclipse, usually with a bright gray or yellow rim to the Earth's shadow.

Four is a bright copper-red or orange eclipse, with a bluish, very bright edge to the Earth's shadow.

If the eclipse seems to be between two categories—one and two, for example—it can be recorded as "1.5 on Danjon's scale." Remember, the scale is for unaided-eye estimates; binoculars or telescopes make the eclipsed Moon appear brighter.

Large telescopes or high magnifications are of little value for observing a lunar eclipse. I recommend using binoculars or a small low-power telescope, since they both allow the entire Moon to be viewed during the event.

During a total eclipse of the Moon, the range of shading and the intensity of color vary enormously. In this case, the core of the Earth's shadow is at upper left, while the shadow's edge is just off the lunar disk at lower right.

CHAPTER TEN

COMETS, METEORS AND AURORAS

I have watched a dozen comets, hitherto unknown, slowly creep across the sky as each one signed its sweeping flourish in the guest book of the sun.

Leslie C. Peltier

On the evening of July 22, 1995, a group of five amateur astronomers from Phoenix, Arizona, decided to head to their favorite observing site south of the city, well away from city lights. The evening began as usual, with the unloading of telescopes, tripods, folding chairs and star atlases. As darkness fell, the telescopes were aimed at galaxies, nebulas and star clusters, the standard fare for a night of recreational astronomy.

At 11 p.m., one of the observers, Jim Stevens, turned his telescope toward the globular star cluster M70, in Sagittarius. After taking a look himself, Stevens then stepped back to allow fellow enthusiast Tom Bopp to have a look. Bopp, who didn't own a telescope at the time, had been invited to join the expedition to share the celestial sights.

As he peered through the eyepiece, he saw the misty glow of M70. At a distance of 65,000 light-years, the cluster's individual stars were delicate specks, like tiny fireflies frozen in flight around a streetlight shrouded in mist. Then Bopp asked, "What's that small, fuzzy object near the edge of the field of view?"

Stevens returned to the eyepiece for a careful look. He saw it too. Assuming that it was a galaxy or a faint cluster which he had never noticed before, Stevens checked his star atlas. According to the atlas, nothing was supposed to be there. Now all the telescopes in the group were trained on the mystery object. There was no doubt it was real—everyone agreed—but what was it?

By midnight, they were convinced that the object had moved, if only slightly. That could mean just one thing: Comet!

Tom Bopp jumped into his car and rushed home to file his report by E-mail to the International Astronomical Union's headquarters. Meanwhile, in neighboring New Mexico, comet hunter Alan Hale was taking a break from scouring the sky for undiscovered comets. He had decided to look at some of his favorite cosmic showpieces,

Comet Hale-Bopp, the brightest comet visible in the evening sky in the northern hemisphere since Halley's Comet in 1910, sported two magnificent tails. The almost featureless white tail is dust illuminated by sunlight. The blue tail is gas excited to luminescence by the solar wind.

next day, they learned that they had discovered a comet, to be known thereafter as Comet Hale-Bopp.

This was no ordinary comet. In late March and early April 1997, when it was nearest Earth, Comet Hale-Bopp became brighter than anything else in the night sky except the Moon and the star Sirius. An estimated one billion of this planet's inhabitants saw it. Not since the historic visit of Halley's Comet in 1910 has such a brilliant comet been visible well above the horizon during midevening from Canada and the United States. Given the rarity of comets of this brilliance, Hale-Bopp should prove to be the brightest comet most people alive today will see during their lifetimes.

Comets are essentially flying mountains of ice left over from the formation of the giant planets Uranus and Neptune. Today, billions of comets are housed in a celestial deep-freeze beyond the orbits of Neptune and Pluto. They orbit the Sun in vast looping paths, taking centuries or even millennia to complete a single trip. Occasionally, slight gravitational nudges from Neptune or a star or nebula passing near the solar system can perturb one of these celestial icebergs in its frigid lair and set it on a course toward the inner solar system, where Earth resides.

As it nears the Sun, the cometary ice begins to be vaporized by sunlight. In the vacuum of space, the vapors create a huge cloud of gas and dust that is many times the

including M70. When he saw the small puffball floating nearby, he knew almost immediately that it had to be a comet. He, too, reported the find, apparently just a few minutes before Bopp. The

Bright enough to be an attention-grabber in moonlight, Comet Hale-Bopp, above, was first magnitude or brighter throughout most of March and April 1997. Hale-Bopp holds the all-time record as the only comet in history visible to the naked eye for more than a full year. Comet Hyakutake, left, seen in March 1996, displayed a longer tail than Hale-Bopp but was less conspicuous overall to the casual observer.

size of Earth. The pressure of sunlight and a constant outward flow of electrons and protons known as the solar wind then push the cloud back into the classic comet tail—actually two tails, a white dust tail and a blue gas, or plasma, tail that shows up more clearly in color photographs than it does to the eye. These tails can be millions of kilometers long.

When at their best in early April, Hale-Bopp's twin tails each spanned a colossal 100 million kilometers of space, more than 200 times the distance from Earth to the Moon. Yet for something so big, the tails contain surprisingly little material, roughly the equivalent amount of gas and dust that would be made by the vaporization of less than one centimeter of snow covering the state of Delaware. A comet's tail is so diffuse that most of it would make a good laboratory vacuum. Yet from afar, seen against the blackness of deep space, comet tails masquerade as objects of substance. Comet expert Fred Whipple once quipped, "Comet tails are the closest thing to nothing that is still something."

The dust tail is usually the brighter of a comet's two tails. Dust is embedded within the nucleus's ice, mostly in the form of particles about the size of window-sill dust. As sunlight vaporizes the outer surface of the nucleus, dust is released with the gas. Just as dust floating in a darkened room is easy to see

Comet Stuff

• Comets are named after their discoverers. The first two people to report a suspected new comet to the International Astronomical Union's headquarters in Cambridge, Massachusetts, are credited with the discovery. About half of all comet discoveries are made by amateur astronomers, usually after years of dedicated searching.

• Halley's Comet is a rare exception to the naming rule. Edmond Halley did not discover the comet but successfully predicted that it would return in 1758, 76 years after it was previously seen. When it last visited the inner solar system in 1986, Halley's Comet was faint (magnitude 4) and inconspicuous from the northern hemisphere. On its previous visit in 1910, it was close enough to Earth to put on a show similar to Comet Hale-Bopp's display in 1997.

• Comet Hale-Bopp is believed to be the biggest comet to visit the inner solar system in more than 200 years. The comet's 35-kilometer-diameter icy nucleus is at least 10 times more massive than the nucleus of Halley's Comet, one of the largest-known comets. Comet Hyakutake, which was visible in late March 1996, passed 13 times closer to Earth than Hale-Bopp but was less than 1 percent of its size.

• Even though Comet Hale-Bopp's nucleus was shedding up to 100 tons of gas and vapor per second when it was at its brightest, less than one-hundredth of 1 percent of the mass of the comet's nucleus was vaporized during its trip through the inner solar system.

• The swept-back tails of comets make them look as if they are moving headfirst, tail last. But the comet's motion has little to do with the tail's streamlined looks. Since solar radiation pushes back the tail, comets always point toward the Sun. As they follow their elongated orbits around the Sun, comets come in headfirst and leave tailfirst.

• Comet Hale-Bopp's orbit is huge. The comet's last pass through our sector of the solar system occurred 4,206 years ago. But thanks to a swing to within 120 million kilometers of Jupiter in July 1996, the comet's orbit was reshaped by the giant planet's gravitational influence, and the comet will now return to the inner solar system in 2,380 years.

A close-up of Comet Hale-Bopp's coma, above, reveals shells of gas and dust thrown off the comet's rotating nucleus. These impressive features were readily visible to backyard astronomers using small telescopes. Left: Close-up of Comet Hyakutake.

Famous & Infamous Comets

Halley's Comet is by far the best-known comet both because of its brightness and because its 76-year orbital period matches the average human life span, so almost everyone gets to see it once and regale their children and grandchildren with tales of the glorious sight. However, this cycle was broken in 1986, when the famous comet was positioned very unfavorably for northern-hemisphere viewers, and it reached only fourth magnitude at its best. By contrast, the big comet's 1910 visit caused a sensation, with its magnificent tail and first-magnitude coma. Astronomers calculate that in the year 837, Halley's Comet passed so close to Earth, it equaled the brightness of the crescent Moon, making it the brightest comet seen by Earthlings in the past 2,000 years.

Comet Shoemaker-Levy 9, discovered in March 1993 orbiting Jupiter, was unique in the annals of astronomy. The giant planet's immense gravitational pull had not only captured the comet but ripped it into 21 pieces. The resulting tribe of mini-comets then crashed into Jupiter in July 1994. The explosions from the impacts produced huge black bruises in Jupiter's atmosphere—some as big as Earth—that were visible in telescopes as small as 60mm refractors. This is the only time humans have witnessed a substantial celestial object colliding with any body in the solar system.

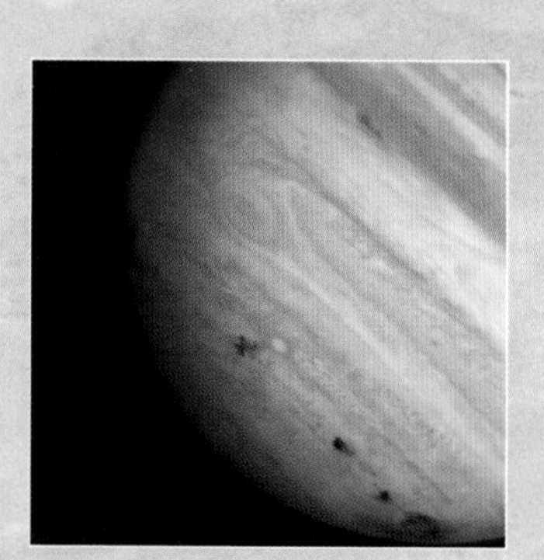

Comet Hyakutake was a small comet that reached first magnitude as it came within 16 million kilometers of Earth in late March 1996. It glided across the sky from Bootes to Polaris in less than a week, brandishing a thin blue tail that reached a length of 65 degrees. It would have remained the brightest comet of the past two decades of the 20th century had not the even brighter **Comet Hale-Bopp** arrived a year later.

In March 1976, **Comet West** was brighter than Hale-Bopp, but only for a couple of days in early-morning twilight. It was not well publicized, and few people apart from aficionados saw it. Other bright comets of the second half of the 20th century include **Comet Bennett** of 1970, **Comet Ikeya-Seki** of 1965 and **Comet Mrkos** and **Comet Arend-Roland**, both seen in 1957.

Some comets were predicted to be spectacular but fizzled. The most notorious of these was **Comet Kohoutek**, which was less than one-tenth of 1 percent as bright as astronomers expected when it reached its peak (magnitude 4) in January 1974. Similarly, **Comet Austin**, hailed in advance as a "monster comet" by the leading astronomy magazines, wimped out and was barely visible to the naked eye in August 1982.

Comet Shoemaker-Levy 9 itself was not visible in backyard astronomers' telescopes, but when fragments of the comet struck Jupiter in July 1994, the impact scars were visible in small telescopes, above. The magnificent curving tail of Comet West, top, graced the early-morning sky over the northern hemisphere in March 1970. Comet Hale-Bopp, right, was bright enough to be seen from the city, where millions of people had their first view of a comet. Photograph shows crowds at the Calgary planetarium.

in a shaft of sunlight, the dust in a comet's tail is a spectacularly effective reflector of sunlight. In fact, some household dust undoubtedly *is* comet dust. As comets shed material into their tails, the dust litters the solar system and is later swept up as the planets orbit the Sun. Larger pieces burn up in the Earth's atmosphere as meteors ("shooting stars"), but the small stuff simply falls to Earth at the rate of a few tons per day worldwide.

Visually, comets have an almost starlike core surrounded by a misty haze called the coma. The diaphanous tail sweeps back from the coma, is usually noticeably fainter than the coma and shows little structure. Occasionally, delicate lengthwise striations make the tail look like strands of hair or give it a subtle feathery appearance. The word comet is derived from the ancient Greek word *kometes*, which means "wearing long hair."

Comets vary greatly in the shape and length of the dust tail and in the brightness of the coma. Some have bright comas and faint stubby tails, some have fanlike tails, and others eject long pencil-thin tails. Frequently, the tail is curved, due to the comet's motion in its orbit. The gas tail, however, is always straight, since it is directly influenced by the solar wind, electrically charged particles streaming outward from the Sun.

When a comet rises above fifth magnitude, usually when it is within Mars' orbit, every effort should be made to observe it each clear night (or early morning, since half of the comets appear in the morning sky). Dark skies substantially enhance the detail visible in a comet's delicate tail structure—especially its overall length. Binoculars give the best views of brighter comets, since the tails usually extend several degrees.

Displaying its 20-million-kilometer-long tail, Comet Hyakutake was near Polaris, the North Star, when this photograph was taken from the author's backyard at 4:05 a.m., EST, March 27, 1996. During the 75-second exposure, the space shuttle Atlantis, docked to the Mir space station, emerged from the Earth's shadow, moving left to right.

Telescopically, a comet nucleus is generally seen as a starlike point, but it is sometimes so shrouded by the coma that it appears simply as a brightened concentration in a haze, like a distant streetlight in mist. Frequently, no structure is detected, especially in fainter comets. The brightest comets show the most activity, and the structural details around their comas are often seen better visually than in photographs. Just as interesting as the specific details of the comet are the changes in size, shape and brightness from night to night as it slowly shifts its position against the starry background.

Meteors

It happens to all of us at one time or another: one quick glance up at the star-filled sky—perhaps for just a second—and suddenly, the placid view is sliced by the brilliance of a falling star. The expression "falling star" is just a description of what appears to be occurring; the object that flashes across the sky and quickly disappears is properly called a meteor.

Meteors have nothing to do with stars. They are tiny bits of space debris so small that thousands would easily fit in your hand. Yet each one of them causes that familiar brief but brilliant flash in the night sky, a dazzling flameout as it ends its existence in a 60,000-kilometer-per-hour plunge into the upper layers of the Earth's atmosphere. At such velocities, friction with air particles vaporizes an average-sized meteor in less than a second. The sudden flash of light is caused by the intense heat of vaporization.

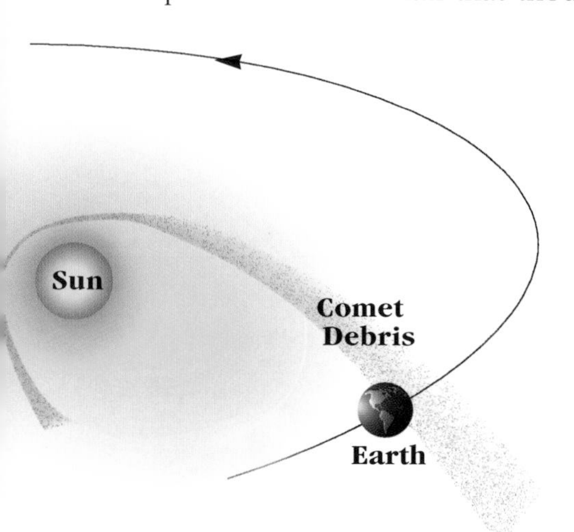

Each day, Earth collects about 400 tons of meteoric debris, most of it microscopic dust so small that it does not produce visible meteors but merely collides with the Earth's atmosphere and floats to the ground months or years later. A minority of the impacting pieces are big enough to flash as visible meteors. On rare occasions, a chunk large enough to survive the fiery plunge hits the ground as a meteorite.

A meteor shower occurs on the same night each year as Earth passes through the stream of debris shed by a comet, above. In some cases, the comet no longer exists, but its dusty trail can linger for millennia. During a meteor shower, the meteors appear to emerge from a point in the sky called the radiant, top. This is a perspective effect; the meteors are actually streaking parallel to each other. Right: A brilliant meteor shares the field of view with the star cluster M7.

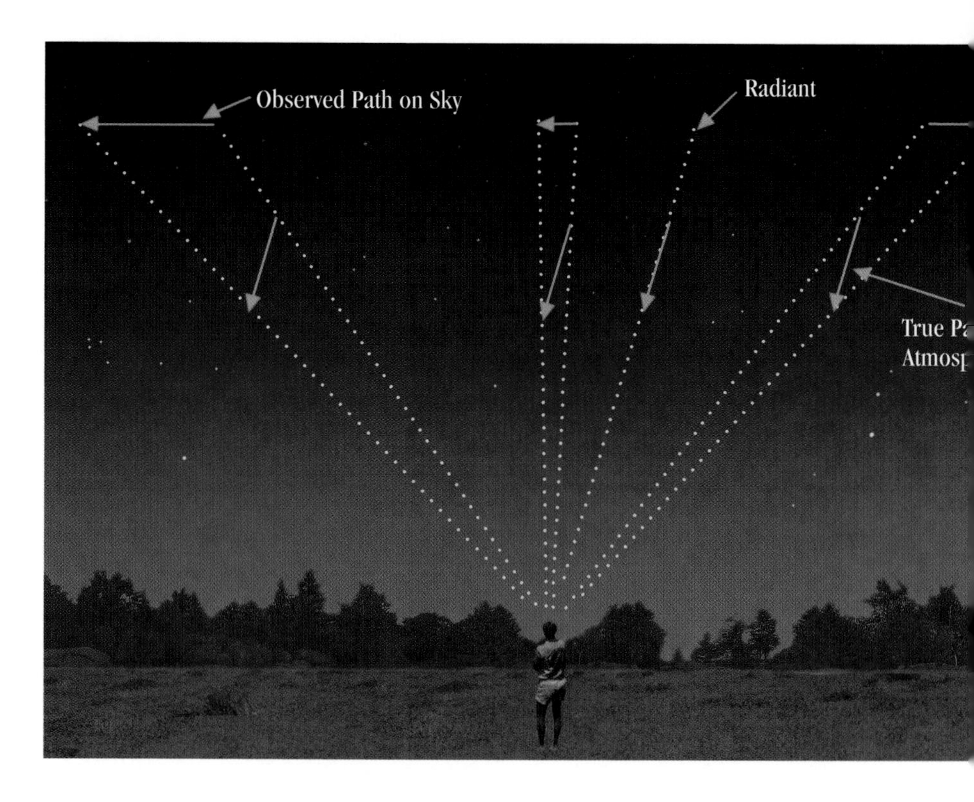

Meteor terminology is cumbersome and confusing, but some of the words need definition. A *meteor* is the bright streak of light seen at night when a small bit of space debris burns up in the Earth's atmosphere. A *meteoroid* is a small chunk of matter in space that could become a meteor. A *meteorite* is a piece of debris that survives its descent through the Earth's atmosphere and reaches the surface. The related term *asteroid* refers to small subplanetary bodies that orbit the Sun, generally between the orbits of Mars and Jupiter. Large meteorites are believed to be debris from collisions between asteroids.

Regardless of the confusing terms, the objects seen in the night sky are meteors. On an average clear night at a dark site, three or four moderately bright meteors per hour can be seen, with the rate rising to seven or eight per hour by dawn. However, as Earth navigates its annual orbit around the Sun, it encounters swarms of meteoric material at predictable intervals. Located at specific points in the Earth's orbit, like mileage posts on a racetrack, the meteor swarms are rapidly encountered and soon left behind. The result is a meteor shower lasting a few nights at most. The best meteor showers can produce one meteor per minute, on average, although most yield fewer.

Whether a meteor shower is occurring or not, the peak period for meteors is from about 1 a.m. to dawn, because after midnight, the nighttime side of Earth faces in the direction in which it is moving in its orbit around the Sun. The after-midnight, or "forward," side of Earth sweeps up more meteors than the before-midnight, or "trailing," side. As an analogy, when I am out walking during a heavy snowfall, the front of my coat becomes plastered with snow, while my back is only slightly peppered with flakes. This, of course, is due to forward motion—I walked into the flakes. Earth does the same as it pursues its orbit at a constant 108,000 kilometers per hour.

Experienced meteor watchers use padded lawn chairs that adjust to a nearly horizontal position so that as much of the sky as possible can be comfortably viewed. Select the darkest available site, and face in the direction of the meteor radiant (see table below). Remember to have a good supply of blankets and insect repellent. Binoculars or telescopes are useless for observing meteors, because their field of view is far smaller than that of the human eye. Meteors can dart unpredictably from practically anywhere. Since meteor observing requires no special equipment, it is a unique opportunity to introduce astronomy to others and to become reacquainted with the stars and constellations.

Serious meteor watchers record each meteor's track on a photocopy of an all-sky chart, such as those in this book. This requires a heavily filtered red flashlight so that night vision will not be spoiled. When a meteor is spotted, its starting and ending points among the stars are carefully noted, then the path is plotted on the chart, with an arrow indicating the direction of flight. Tracing the lines backward at the end of the night reveals the radiant point of the meteor shower. A few meteors will not radiate from this area. These are the random nonshower meteors that can be seen any night of the year.

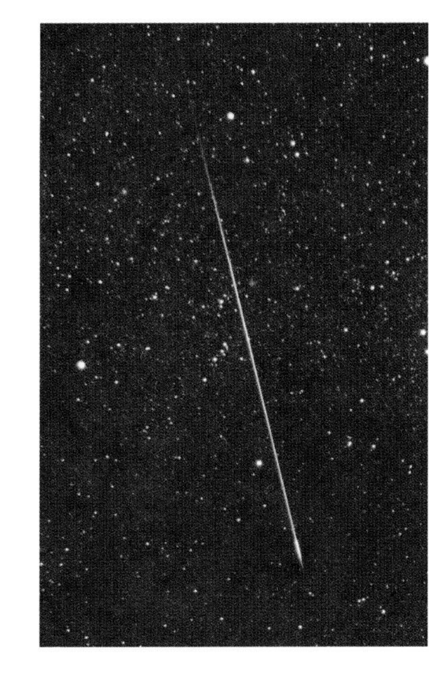

The best way to capture a meteor on film is to let the stars trail during a several-minute exposure on the peak night of a meteor shower, top right. Above: A 1997 Leonid meteor slashes through the constellation Auriga.

Meteors from the recognized showers are known to be debris from comets. The comets relinquish the meteoric material during close encounters with the Sun. Their icy bodies partly vaporize, releasing embedded dust and denser bits of ice, which spread along the comet's orbit like a trail of sand from a punctured sandbag. When Earth crosses this trail of debris, a meteor shower occurs.

Even during the most abundant of the regular meteor showers, when several dozen meteors per hour may be observed under favorable circumstances, the average distance between individual meteors is more than 100 kilometers, so the "swarm" that produces the shower is, in fact, almost empty space. No meteor belonging to any meteor shower has ever been known to reach the Earth's surface. Even those as bright as Venus are quickly consumed, but they often leave a glowing trail that may last for several seconds.

Major Annual Meteor Showers

Name of Shower	Radiant	Date of Maximum	Hourly Rate at Maximum*
Quadrantid	NE (Draco)	Jan. 3	10-50
Lyrid	NE (Lyra)	Apr. 21	5-25
Eta Aquarid	E (Aquarius)	May 4	5-20
South Delta Aquarid	SE (Aquarius)	Jul. 27-29	10-20
Perseid	NE (Perseus)	Aug. 12	30-70
Orionid	E (Orion)	Oct. 20	10-30
Leonid	E (Leo)	Nov. 16	10-20†
Geminid	E (Gemini)	Dec. 13	30-80

*The range of values reflects the variation in the strength of the meteor displays from year to year. These figures do not include the half-dozen or so sporadic meteors seen each hour.
†Leonids expected to be stronger in 1998 and 1999.

Auroras

Diaphanous curtains of green, white and red dance in the northern sky, billowing and swirling as if propelled by some distant cosmic wind. Most of us have seen this nocturnal spectacle at one time or another—the aurora borealis, or northern lights. Some nights, these displays look more like pulsing clouds or filmy arcs of light on the northern horizon. Or, more rarely, the sky is alive with roiling luminescence overhead.

From southern Canada and the northern plains of the United States, several impressive auroral displays are seen annually, on average. And once or twice a decade, the auroral lights dance as far south as Florida.

The most persistent (and entirely wrong) explanation for the phenomenon is sunlight reflecting off polar ice. As long ago as the early 1900s, Norwegian astrophysicist Carl Stormer correctly explained that auroras originate with energetic particles blasted in the Earth's direction by eruptions on the Sun. Traveling at speeds of millions of kilometers per hour, these particles reach Earth in a day or two, but instead of plunging directly into the atmosphere, they are deflected by the planet's magnetic field.

The most frequent and intense auroras are seen from Alaska or the Canadian Arctic, where the above photograph was taken. But a few times every decade, immense displays blanket most of the continent, such as the night of March 12-13, 1989, when the photograph at top was taken from the author's backyard. Crimson auroral sheets are typical of these brilliant outbreaks of nature's nocturnal lights.

The particles then funnel along the Earth's magnetic lines of force, eventually entering the atmosphere over the magnetic poles, where they electrically stimulate the rarefied gases of the Earth's upper atmosphere. These gases effectively act like a giant television screen, glowing when bombarded by the particles. Oxygen emits a greenish white light or a red hue at extremely low atmospheric pressures; ionized molecular nitrogen produces a bluish tinge. Auroral displays range in altitude from 100 to 1,000 kilometers, although the most spectacular displays are near the lower end of that scale, where the atmosphere is at the right density to produce the maximum effect.

Because of the direct relationship between auroras and solar activity, periods of high solar turbulence near sunspot maximum always produce increased auroral activity. At sunspot maximum, auroras are more likely to be seen farther south, because the magnetosphere—the cometlike bubble of magnetically trapped particles around Earth—is distorted, extending the auroral zone toward the equator enough to allow millions more people to see the dancing lights.

Records of aurora borealis observations can be traced as far back as the sixth century B.C. Chinese scholars described the lights as a fire dragon among the stars. The most abundant auroral displays recorded in China occurred in the 11th and 12th centuries A.D., coinciding with high levels of solar activity. Greek accounts by Anaximenes and others leave no doubt that the phenomenon has always been a source of wonder.

A typical auroral display begins with a white or pale greenish glow toward the northern horizon. Then a few spikes or arcs slowly crawl up the sky, brightening as they rise. As the aurora increases in intensity, vertical bands begin to shimmer and waver, developing into waves and folds that form fragile celestial draperies. In the best displays, pulsating curtains in hues of red, green and violet can last from a few minutes to hours.

In an intense display, the undulating veils are overpowered by swirling, billowing clouds that lurch over the sky in a matter of seconds. The very best auroras develop into brilliant streamers radiating from overhead, and the sky fills with light and color, obliterating almost all the stars. Called coronal auroras, these are one of nature's greatest spectacles. During a coronal aurora, the rays are actually parallel, but they appear to converge at the magnetic zenith, just as railroad tracks seem to converge in the distance. The magnetic zenith for most of the populated parts of Canada and most of the United States is slightly south of the overhead point. To be fully appreciated, auroras must be observed from a dark location. Their delicate structure and hues are wiped out by streetlights, backyard illumination or even the glow from urban lighting in general.

Reports that people are able to hear auroras are so persistent that many scientists are becoming convinced that the phenomenon is real. There are hundreds of accounts of crackling, hissing sounds associated with auroras, but only some people hear them. In a group, usually one or two people note the sounds. The origin of such sounds is still debated, but it illustrates the fact that cosmic mysteries are as near as the top of the Earth's atmosphere.

Above: Comet Hale-Bopp, near the treetops at lower left, is dwarfed by the streamers of a brilliant aurora on April 10, 1997. The 30-second exposure was taken using the setup shown at top left on page 166. Right: A classic green-hued aurora.

CHAPTER ELEVEN

PHOTOGRAPHING THE NIGHT SKY

Astronomy offers one of those pleasures which follow the law of increasing, rather than diminishing, returns. The more you develop it, the more you enjoy it.

Viscount Grey

With the exception of a few spacecraft images, virtually all the astronomical photographs in this book were taken by amateur astronomers. Some of the pictures were simple to obtain; others required specialized equipment, hours of preparation and plenty of skill and experience on the part of the photographer. This chapter is a step-by-step guide to the easiest types of astrophotography.

First, the camera. Since the vast majority of successful astrophotos are taken with 35mm single-lens-reflex (SLR) cameras, you really should have one even for dabbling in celestial photography. The camera must allow the shutter to remain open for a time exposure of at least 30 seconds—preferably indefinitely. Most 35mm SLR cameras have this option, although many other types, such as point-and-shoot cameras, do not. Look for a setting marked "B" (meaning bulb). A plain old manual 35mm camera such as an Olympus OM-1 or a Pentax K-1000 is ideal. Both are out of production but are still available in camera stores that sell used equipment.

Next, the film. Use color. Black-and-white films have no superiority except in advanced applications well beyond what beginners should try. There is more color in the sky than the eye can see, and modern color films pick it up easily. This is one area where recent advances in film technology work completely to the benefit of amateur astronomers. New films like Kodak Royal Gold 1000 and Pro 1000 and Fujicolor 400 and 800 are vastly improved for night photography over anything available previously. These are my favorite films, but any print or slide film rated at 400-speed or higher will work well.

With slide film, your "negative" is your slide. What you see projected on your home screen is what you shot. If you goofed, it is apparent. Color print films are different in that the rectangular picture you get back from the photofinisher is a printed version of

This 20-second exposure on 100-speed film captured Comet Hale-Bopp, the crescent Moon (overexposed) and the Pleiades cluster in between. The tripod-mounted 35mm camera had a 24mm f2.8 lens.

the negative. Photofinishers (whether machine or human) are used to printing pictures of kids, pets and vacation scenery. Astronomical photographs often throw them. Sky backgrounds can appear as anything from black to green to pink. Seldom do you get the best print that can be made from your negative. What to do? Tell your photo dealer that there are "night-sky pictures with stars" on the film. This will alert the film lab to something unusual and reduce the chances of wretched prints.

If you think you have captured something good, ask for a custom reprint. Although these prints cost more, you can specify exactly what you want, and the film lab is obliged to do it over again, if necessary, to get it right. Always take a sample of some good astrophotos to give the photo technician an idea of what you're looking for.

Most 35mm cameras have interchangeable lenses. This allows both wide-angle and telephoto possibilities. Zoom lenses have become popular in recent years. If zooms are all you have, use them. However, fixed-focal-length lenses are preferred for astronomical photography because they have "faster" focal ratios around f2.8 and usually produce sharper images. Now, on with how you can take a celestial photograph.

For the easiest type of sky photography, you will need a 35mm camera with a 55mm or shorter focal-length lens, a tripod and a cable shutter release. The cable release (less than $10) is necessary to lock the shutter open for time exposures. Load your 35mm camera with 400-speed or faster film, mount it on the tripod, turn off anything automatic that can be turned off (light meter, flash, et cetera), and set the lens at f2.8 or to the lowest f-setting. Focus at infinity. Set the shutter speed at B, and aim the camera at a familiar stellar configuration, say, Orion or the Big Dipper. Place the lens cap over the lens. Press the cable release, and lock the shutter open with the lock on the cable.

Now you are ready for the picture. Gently remove the lens cover for 12 to 25 seconds with a 50mm lens, 20 to 35 seconds with a 28mm or 24mm lens. Replace the cover, and release the cable lock. Using this technique, you will get realistic pictures of the night sky showing lots of stars—and this is absolutely the best way to capture an aurora on film. Stars well below the naked-eye limit will be recorded if conditions are good. Take several exposures within the suggested time limits. The Earth's rotation causes exposures near the maximum times to produce tiny trails for the stars instead of dots. If the exposure is longer still, the trails become obvious.

Use the Earth's rotation to your advantage when shooting the night sky. Capturing this motion with tripod-mounted-camera exposures of many minutes—even hours—results in majestic portraits of sweeping stellar arcs. Set up the camera as described, but leave the shutter open for at least 15 minutes. Such a lengthy exposure relaxes the film choice (see box, facing page). Photographs taken with the camera pointing toward the southern sky or overhead will show long east-to-west trails, whereas photographs taken pointing toward the north pole of the sky (an imaginary point very close to Polaris, the North Star) produce concentric star trails with the geometrical center of each curving arc at the north celestial pole. An exposure of an hour or more will show beautiful swirling trails as the stars appear to pivot around Polaris, as seen in the photograph on page 28.

The star colors recorded by color film are true colors. Each star is a slightly different temperature. Hot stars are blue, cool stars are white, and even cooler stars are shades of yellow and orange. Occasionally, an aircraft or a satellite sweeps across the camera's field of view during star-trail photography. In both cases, the star trails will be straight, but the aircraft's flashing lights may produce an

A standard tripod-mounted 35mm camera, top, was used for the picture on page 165 as well as many others in this book. A cable release, seen on every camera shown in this chapter, allows the photographer to begin and end the exposure without jiggling the camera. Equatorial mounts with built-in polar-axis alignment scopes, such as the one at right, made by Vixen, are ideal for tracked long-exposure shots. Lower inset shows the view through the alignment scope. To achieve near-perfect polar alignment in the northern hemisphere, the mount is adjusted to place Polaris in the small offset circle and the star Delta, in Ursa Minor, in another circle (not shown) at the end of the line extending to the left in this view. The setup takes 5 to 10 minutes.

CAMERA ON TRIPOD

TRACKED BY EQUATORIAL MOUNT

Sky Portraits
To capture portraits of the night sky approximately as the eye sees it, use wide-angle lenses, such as 28mm and 24mm, that allow plenty of opportunity to frame the scene with a familiar foreground. Exposures with these lenses should be in the range of 20 to 40 seconds to avoid significant star trailing. Use 400-to-1600-speed film (800- and 1000-speed films are especially recommended), and set the lens at its lowest f-stop. This is the ideal setup to record constellations and auroras.

Star Trails
Dark skies and long exposures are the key to shooting impressive star trails. For a 1-hour exposure, set the lens f-stop to f2.8 with 100- or 200-speed film, f3.5 or f4 with 400-speed. For longer exposures, use slower f-stops (higher numbers). In the light of a gibbous or full Moon, close down the aperture by at least two additional f-stops. Moonlight will illuminate the Earthscape to near-daylight level. Any focal-length lens will work for star trails, but 28mm and 24mm are the most versatile.

Horizon + Deep Sky
Attaching your camera to a polar-aligned motor-driven equatorial mount, such as the one pictured on facing page, keeps the stars from trailing, but the ground will trail, because the camera is following the sky's motion. If you keep the exposures to 1 to 2 minutes, the foreground is only slightly blurred while impressive detail is recorded in the sky. (The picture above was 100 seconds at f2 on 800-speed film, 24mm lens.) Again, 28mm and 24mm lenses are favored for their generous fields.

Deep-Sky Exposures
Extending the equatorial mount's guiding for 5 to 10 minutes results in tremendous detail using 400-to-1000-speed film. The image above, the region around the North America Nebula in Cygnus, is a section of a 10-minute exposure on 400-speed film taken with a 180mm f2.5 lens. Accurate polar alignment is necessary to avoid star trailing. This technique produces impressive results with lenses up to 200mm focal length. Longer lenses require more advanced guiding techniques.

interesting beads-on-a-string effect. Another unpredictable intruder, and by far the most prized unintentional catch, is a meteor slashing across the star trails. Only bright meteors will show up; the faint ones are too feeble to register on film during their brief flights.

On the peak nights of the best showers, such as the Perseids or Geminids, try meteor hunting with a camera by setting up for star-trail photographs. Aim the camera in the general direction of the radiant (see table in Chapter 10). It is largely a matter of luck, but you could bag a real "keeper" using this method.

A telephoto lens of 100mm focal length or longer expands the repertoire of available celestial objects to include the Moon. Some of the most beautiful pictures of our satellite are taken just before the sky is completely dark, with the Moon forming a backdrop for trees or houses or other foreground objects. (Lenses of less than 100mm focal length do not show the Moon large enough to distinguish any of the darker surface areas that characterize its features.) Telephoto shots of the Moon are best regarded as trial-and-error efforts, although modern light-metering systems can be fairly accurate. Exposures are generally less than 1 second, sometimes as little as 1/250 second, but the important point is to take a range of exposures and to record the pertinent data so that successes can be repeated. A camera tripod is usually necessary.

A rewarding and relatively uncomplicated extension of camera-on-tripod astrophotography is guided photography using a motor-driven equatorial mount. The photographs at

Fully manual used cameras, such as the Olympus OM-1, facing page, center, are still widely available and make ideal astrocameras. Viewscreen magnifiers, such as the 6x unit on the Nikon at right, are prized for astronomical use.

The Barn-Door Tracker

An equatorial mount is not the only way to take guided astrophotos. An inexpensive alternative, known as a barn-door tracker, can be put together on a Sunday afternoon. You need no special equipment to build it and no motors or batteries to run it. It's driven by you, turning a hand crank. This is low-tech at its best.

The principle behind the barn-door tracker is that its hinge parallels the Earth's axis. Turning a hand crank that pushes on one of the two hinged boards compensates for the Earth's rotation. The camera, attached to the moving board, tracks the stars.

The most expensive parts of the barn-door tracker are the heavy-duty camera tripod to hold the device and the ball-head camera clamp to hold your camera at any orientation atop the tracker. Other than these two items, the materials are very inexpensive; you may find most of them around the house. Specific details of the design are up to you, but here are the basics.

In the model shown here, two standard door hinges were used, but even a single standard hinge will work. The drive bolt is a 4-inch-long ¼ x 20 carriage bolt. Two ¼ x 20 T-nuts are the only specialized hardware required.

Cut two pieces of ¾-inch plywood into two T-shapes, and join the tops of the T with the hinges. The only crucial measurement is the position of the hole that contains the T-nut, which holds the drive bolt. It should be exactly 11 7/16 inches from the center point of the hinge axis. To secure the unit to a standard camera tripod, the bottom leaf needs a ¼ x 20 threaded hole, like that found on the bottom of cameras.

Attach the other T-nut here, but be sure to countersink the T-nut from the top so that the tripod pulls the T-nut down into the bottom board. A four-pointed handle makes turning the drive bolt at one rotation per minute easy. Adequate accuracy for a typical 5-minute astrophoto with a 50mm or shorter focal-length lens is achieved by *gently* cranking the drive bolt clockwise a quarter-turn every 15 seconds.

To work properly, the axis of the hinges in this design must be aimed at Polaris, the North Star, with the drive bolt on the right as you face north. A small tube parallel to the hinges acts as a sight for the alignment.

lower right, on page 166 show how I do it. I prefer a mount with a polar-alignment scope integrated into the polar axis. These little scopes make accurate polar alignment a snap. Mounts like this are available new or used at many telescope dealers. A ball-head camera mount attached to the equatorial mount offers a quick and secure way to point the camera in any direction. The equatorial mount then compensates for the Earth's rotation, keeping the camera aimed at the same spot in the sky for a long exposure. Star trailing is eliminated, and the longer exposure time will record much fainter stars. Use 50mm or shorter focal-length lenses, at least initially, because they are more tolerant of slight errors in polar alignment.

Now I come to what everybody wants to do—or *thinks* he or she wants to do—astrophotography *through* the telescope. Other than shooting the Moon, this is far more difficult than it seems to the beginner. Yet at the same time, it can be hugely rewarding. It demands a great deal of patience, plenty of trial and error and, in many cases, very dark skies. Here, I must refer you to other sources for details, in particular *Splendors of the Universe*, which I coauthored with astrophoto guru Jack Newton, and *The Backyard Astronomer's Guide*, coauthored with Alan Dyer, another highly respected astrophotographer.

But let's return to shooting the Moon, which is something that can be done with almost any telescope. The vital accessory is a camera/telescope adapter (about $50). It replaces the lens on an SLR camera with a tubular holder the same size as a telescope eyepiece. The camera then replaces the eyepiece and uses the telescope as a telephoto lens. This method is almost foolproof for lunar photography and also works well in full daylight on terrestrial subjects. The main thing to watch for is vibration while taking the photograph. The shutter "slap" on a typical 35mm SLR can introduce just enough vibration to blur the picture. There are detailed exposure

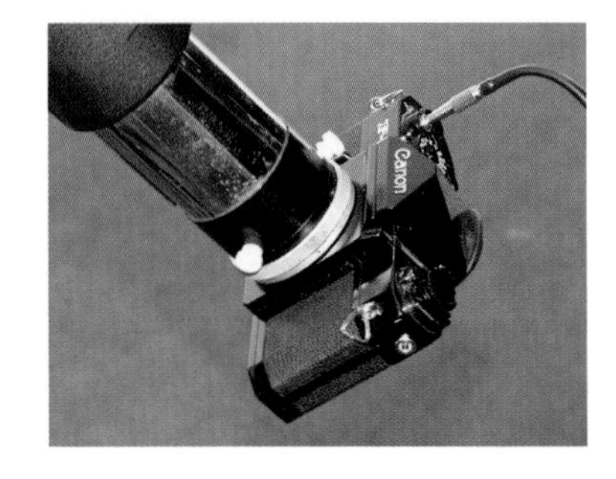

Attaching a camera to a telescope requires two accessories: a T-ring to fit your camera (available from any camera store) and a 1.25- or 2-inch camera/telescope adapter (available only at telescope stores). These two items are joined together, top right, and attached to the camera, top left. The camera is then fitted like an eyepiece at the telescope's focus, above. In this case, a 2-inch adapter was used. With such a setup on a solid mount, you should get a decent Moon portrait on your first try.

tables for lunar photography in the two books mentioned above, but to begin, try 1/30 second with 100-speed film on the first-quarter Moon, and bracket either way from there.

Photographing the planets is *much* more difficult. For the most part, the results are disappointing, because even the best planetary photographs never show as much detail as can be seen visually through the eyepiece of the same telescope. Planetary photography also requires additional adapters to place an eyepiece ahead of the camera to boost the image to a suitable size. Planetary photographs or extreme close-ups of lunar features need exposures of up to 15 seconds. Because of these drawbacks, almost all planetary photographers have converted from film to CCD (electronic) imagers, which require shorter exposures and are more sensitive to subtle detail.

The ultimate challenge for the astrophotographer is the unlimited number of galaxies, nebulas and star clusters beyond the solar system. To shoot them properly, an accessory called an off-axis guider or a separate guiding telescope is needed to keep the telescope precisely aimed during the exposure. This involves motor-driven slow-motion controls on the telescope and a push-button control paddle or autoguider to keep a guide star centered while shooting the main object. To do it right requires a considerable investment in accessories and more instructions than can be provided here. For the most up-to-date developments in this field, check the astrophotography articles in *Sky & Telescope* and *Astronomy* magazines over the past few years. The top amateur astrophotographers are producing truly breathtaking images, an amazing testament to their dedication and ingenuity.

CCD Cameras: Astro-Imaging Revolution

Astronomical CCD cameras are like video camcorders except that they are designed to take single exposures ranging from a fraction of a second to many minutes in length. In a CCD camera (and a video camcorder), film is replaced by a silicon chip called a charge-coupled device, hence the name CCD. Most CCD chips measure only millimeters across, usually much smaller than a frame of 35mm film, but they contain hundreds of thousands of tiny square pixels. Each light-sensitive pixel records the amount of light that strikes it during an exposure and converts that reading into an electronic signal. At the end of an exposure, the signals from the thousands of pixels are recorded as digital data and stored on a computer's hard drive. The whole picture can then be displayed on a monitor.

The imaging surface of a CCD is relatively small but incredibly sensitive to light—about 15 times as sensitive as photographic film. CCDs literally count almost every photon of light the telescope delivers. Moreover, their sensitivity does not wane during a time exposure the way film does (a defect called reciprocity failure). These two advantages mean that a 1-minute CCD exposure records about the same detail as a 30-minute exposure on film. A 10-minute CCD image captures much fainter detail than film ever can with the same telescope.

The disadvantages of a CCD include the smaller area of sky recorded compared with photographic film, monochrome rather than color imaging and the fact that a computer is required to store and examine the images. However, all these considerations have their positive side too. The small but detailed image is preferred for some objects (planets and most galaxies, for instance), and color can be obtained by combining filtered images. Finally, computers allow easy storage, access and image processing.

Then why use film? For casual sky shooting, a 35mm camera is already available to most people and is relatively low-tech. For simple constellation or star-trail shots, the only accessory required is a camera tripod. For wide-field detail, such as guided portraits of constellations, bright comets, the Milky Way, larger star clusters, and so on, photographic film's comparatively high resolution over a large frame still has the upper hand over all but very costly research-grade CCDs. The advantages of CCDs (at least at their present level of development) are best applied to objects that appear faint and/or small. Another consideration is that at minimum, a CCD imager and the related image-processing software cost at least $1,000, plus the telescope and computer. However, CCD technology is improving much faster than developments in film, so it may not be too many years into the 21st century before almost all celestial photography will be gathered with CCD cameras.

The heart of a CCD camera is the tiny CCD chip, top left. The CCD camera is inserted into the focuser just like an eyepiece. A computer controls the camera and stores the downloaded images for processing and combining to achieve color.

CHAPTER TWELVE

RESOURCES

What is it all but a trouble of ants in the gleam of a million million suns?

Alfred, Lord Tennyson

Like all leisure interests, astronomy requires a modest initial investment in equipment and supplies. At minimum, you need decent binoculars, a few general astronomy books, one or two practical guidebooks, an annual data-reference book and a subscription to at least one astronomy magazine.

Magazines

There are two major monthly magazines designed for and, in many instances, written by amateur astronomers. Thousands of astronomy enthusiasts subscribe to both, and reading at least one of the two regularly will keep you up to date on celestial events and discoveries. In addition, the two magazines are the best places to find out who sells what astronomy equipment. Both are filled with informative ads.

Sky & Telescope, Box 9111, Belmont, Massachusetts 02178 (subscriptions: 800-253-0245; Website: www.skypub.com/). All serious amateur astronomers, many professionals and thousands of libraries subscribe to this excellent publication. Copies are found on larger newsstands, and it is often sold by planetarium shops and telescope dealers. The magazine attempts to cover the entire field, from recent discoveries in astrophysics to reports on amateur-astronomy conventions. Current sky events are well covered, with excellent diagrams and charts.

Astronomy, Box 1612, Waukesha, Wisconsin 53187 (subscriptions: 800-533-6644; Website: www.astronomy.com). This colorful publication is the world's largest-circulation English-language astronomy magazine. Most well-stocked newsstands carry it. Each issue contains several lavishly illustrated features along with articles on the use of astronomical equipment, observing techniques and current celestial events.

SkyNews, National Museum of Science & Technology Corp., Box 9724, Station T, Ottawa, Ontario K1G 5A3 (subscriptions: 800-267-3999). Canada's astronomy magazine is a full-color bimonthly publication with articles and sky charts aimed at the beginner. It is also available as a benefit of membership in The Royal Astronomical Society of Canada.

The Planetary Report, a bimonthly magazine devoted entirely to planetary exploration and to the search for extraterrestrial life, is available through membership in The Planetary Society, 65 N. Catalina Avenue, Pasadena, California 91106 (subscriptions: 626-793-1675; Website: planetary.org).

The monthly *Abrams Planetarium Sky Calendar* is a concise digest of what is visible with unaided eyes and binoculars. It is not a magazine, just two sides of a standard-sized page per month, but a surprising amount of information is clearly presented in neat diagrams that are useful to beginners and veteran astronomers alike. A year's subscription is just a few dollars. For subscription information, write: Sky Calendar, Abrams Planetarium, Michigan State University, East Lansing, Michigan 48824.

Practical Stargazing Guides

The next step beyond *NightWatch* is *The Backyard Astronomer's Guide* (Firefly Books) by Terence Dickinson and Alan Dyer. It intentionally has little overlap and greatly expands on such topics as observing techniques, telescope selection and astrophotography. Another companion book is *Summer Stargazing* (Firefly Books), which has charts and information focused on the summer sky.

A classic reference book that should be on every astronomer's shelf is *Burnham's Celestial Handbook* by Robert Burnham Jr. (Dover), a colossal 2,100-page three-volume set that took literally decades to prepare. Thousands of multiple stars, variable stars, star clusters, galaxies and nebulas visible in amateur astronomers' telescopes are described or tabulated in detail. In softcover, the three books are about $25 each.

My favorite sixth-magnitude star atlas is *Bright Star Atlas* by Wil Tirion (Willmann-Bell); for an excellent free astronomy-book catalog, write: Willmann-Bell Inc., Box 35025, Richmond, Virginia 23235. All stars to magnitude 6.5 are shown, along with thousands of deep-sky objects on tables facing the charts. A bargain at less than $20. A color version ($40) using charts in a slightly different format is *The Cambridge Star Atlas 2000* by Wil Tirion (Cambridge).

A new, updated version of Wil Tirion's *Sky Atlas 2000* (Sky Publishing)—a well-designed, beautifully detailed eighth-magnitude star atlas—was released in 1998. A small section of this fine atlas is reproduced on page 95.

An impressive ninth-magnitude star atlas for

amateur astronomers is *Uranometria 2000* (Willmann-Bell), showing 330,000 stars on 473 charts bound into two volumes (about $100 a set). Virtually every deep-sky object visible in any amateur telescope is plotted and named. A companion book, *The Deep Sky Field Guide* (Willmann-Bell), tabulates data about every charted deep-sky object.

The *Millennium Star Atlas* ($250) is the ultimate printed star atlas. In three huge volumes, its 1,500 charts display enormous detail, with a million stars to magnitude 11. This is not a reference for beginners, but experienced stargazers and astronomy clubs will want to own this definitive work from Sky Publishing.

Some other useful references are: *Binocular Astronomy* by Crossen and Tirion (Willmann-Bell); *Touring the Universe Through Binoculars* by Philip Harrington (Wiley); *The Guide to Amateur Astronomy* by Newton and Teece (Cambridge); *Advanced Skywatching* by Robert Burnham et al. (Time-Life); *The Sky: A User's Guide* by David Levy (Cambridge); *Star-Hopping for Backyard Astronomers* by Alan McRobert (Sky Publishing); *Eclipse!* by Philip Harrington (Wiley); and *The Year-Round Messier Marathon Field Guide* by Harvard Pennington (Willmann-Bell).

Veteran sky observer Fred Schaaf covers many aspects of naked-eye stargazing in *The Starry Room* (Wiley), a series of personal essays containing useful tips and insights.

The Astronomical Companion by Guy Ottewell successfully melds many concepts that are left dangling in introductory astronomy books. Ottewell has crafted a series of innovative illustrations that place celestial phenomena and the large-scale structure of the universe in perspective. I highly recommend it (available only from the publisher: Universal Workshop, Dept. of Physics, Furman University, Greenville, South Carolina 29613).

The best book for the typical first-time telescope maker is Richard Berry's *Build Your Own Telescope* (Kalmbach). Berry is an expert designer of simple, rugged backyard telescopes. For practical astrophotography, try *Splendors of the Universe* by Terence Dickinson and Jack Newton (Firefly Books) and *Astrophotography for the Amateur* by Michael Covington (Cambridge). By far the best guide to the Moon is Antonin Rukl's *Atlas of the Moon* (Kalmbach). Rukl's beautifully rendered maps and concise descriptions of lunar features are superb.

Planetariums

Most major cities in North America have public planetariums that stage "sky shows" under a huge projection dome by creating realistic images of the night sky, as seen at any time on any day from any place on Earth. During the 1930s, when the first big Zeiss planetariums were installed in New York, Chicago and Philadelphia, the shows caused a sensation. Every large city had to have one.

Today, in an era of virtual-reality games and stunning Hollywood visual effects, the planetarium has lost much of its impact. Few new planetariums are being built, and some existing ones have been severely squeezed by budget cuts. However, a positive planetarium experience, especially for a youngster, can still last a lifetime. Because most planetariums produce their own shows, however, the quality of the productions varies from genuinely thrilling to, frankly, appalling. The problem is, once you have seen a boring show, you may not have the desire to go back to *any* planetarium. If this happens, try another planetarium. Sooner or later, you will see how it can and should be done. Or if your local planetarium offers quality productions, support it by seeing every new show.

Some planetariums offer introductory astronomy courses that are well worth taking. Planetariums usually have telescopes available for use during such courses. A planetarium bookstore is often the best source in town for books, charts and reference materials. In many cases, the local astronomy club holds its meetings at the planetarium. If the planetarium staff is doing its job, the building should be the center of astronomical activity for the city and surrounding area.

Constellation Lore

Star Tales by Ian Ridpath (Universe Books) and *The Starlore Handbook* by Geoffrey Cornelius (General Publishing) recount the classic myths associated with each constellation. Less well-known myths of other cultures are included in *The New Patterns in the Sky* by Julius Staal (McDonald & Woodward). A more technical reference is *Star Myths of the Greeks and Romans* by Theony Condos (Phanes Press).

General Astronomy Books

Excellent overviews of current astronomical knowledge are provided by introductory college textbooks such as *Astronomy: The Cosmic Journey* by Hartmann and Impey (Wadsworth Publishing) and *Astronomy: From the Earth to the Universe* by Jay M. Pasachoff (Saunders). These are two of my favorites, but there are dozens more in this category.

Less technical than the above is *The Universe and Beyond* by Terence Dickinson (Firefly Books), which covers the full spectrum of astronomical inquiry, from solar system exploration to cosmology to extraterrestrial life. Three outstanding descriptive astronomy books for general readers are *Coming of Age in the Milky Way* by Timothy Ferris (Morrow), *The Whole Shebang* by Timothy Ferris (Simon & Schuster) and *The Pale Blue Dot* by Carl Sagan (Random House), all of which represent science writing at its finest.

Light Pollution

Although there is as yet no book completely devoted to light pollution, excellent reference material is available from the International Dark-Sky Association, 3225 N. First Avenue, Tucson, Arizona 85719 (520-293-3198; Website: www.darksky.org). Articles on light pollution appeared in the September 1998, May 1998 and November 1996 issues of *Sky & Telescope* magazine. The subject is also covered in *The Backyard Astronomer's Guide*.

Slides, Posters, CD-ROMS

A free catalog of quality astronomical goodies is available from each of these suppliers: Astronomical Society of the Pacific, 390 Ashton Avenue, San Francisco, California 94112 (800-335-2624); Hansen Publications, 1845 South 300 West #A, Salt Lake City, Utah 84101 (800-321-2369); Sky Publishing, Box 9111, Belmont, Massachusetts 02178 (800-253-0245); The Planetary Society, 65 N. Catalina Avenue, Pasadena, California 91106 (800-969-6277; Website: planetary.org).

Annual Publications

The *Observer's Handbook*, published annually by The Royal Astronomical Society of Canada, 136 Dupont Street, Toronto, Ontario M5R 1V2, is an essential reference for any backyard astronomer. Dozens of tables for celestial phenomena range from times of moonrise and moonset to the positions of Jupiter's moons and the location of meteorite scars in North America. Thoroughness and accuracy have made the *Observer's Handbook* the most widely used annual reference for amateur astronomers.

Astronomical Calendar by Guy Ottewell is published by Universal Workshop, Dept. of Physics, Furman University, Greenville, South Carolina 29613. Write for a descriptive brochure and for the current price of this excellent annual guidebook. Ottewell's unique illustrations and emphasis on diagrams, rather than tables, produce little direct overlap with the *Observer's Handbook*. I use both.

Two other annual publications, *Explore the Universe* and *SkyWatch*, are prepared by *Astronomy* and *Sky & Telescope* magazines, respectively. Both are widely available on newsstands during autumn. Though less comprehensive than the *Observer's Handbook* and *Astronomical Calendar*, they contain plenty of useful information for the coming year, and at about $6 each, they are an excellent value.

Computer Software

The evolution of astronomy software for personal computers has been swift and dramatic over the past few years. There are at least 50 different programs of varying capabilities available. The most versatile for the average recreational astronomer are planetarium-type programs that allow the user to select a date, time and location and have the sky realistically portrayed, complete with horizon, planets, stars, constellations (with or without connecting lines), and so on. Many other features are usually included, but easy-to-use planetarium capabilities should be paramount.

Among the most popular of the planetarium-type software is *TheSky* (Software Bisque; Website: www.bisque.com), available in several increasingly sophisticated levels for PCs. Other highly regarded programs are *RedShift* (Maris Multimedia Inc.; Website: www.maris.com) and David Lane's *Earth Centered Universe* (Nova Astronomics; E-mail: info@nova-astro.com; Website: www.nova-astro.com).

Two superb programs for Macintosh computers are *Voyager* (Carina Software; Website: www.carinasoft.com/) and *Starry Night* (Sienna Software; Website: www.siennasoft.com). In 1998, *Starry Night* also became available for Windows 95/NT operating systems.

Some Useful Websites

Here are a few general Websites on the Internet. The list is rather short because a number of other Websites have already been mentioned in this chapter. Also, most astronomy buffs quickly develop specific interests that are catered to by more specialized sites.

nasa.gov/today/ NASA posts news releases and announcements on this news page. Every news story issued by NASA in the past decade can be found here. Not long ago, in the days of snail-mail, this material was distributed only to journalists.

science.msfc.nasa.gov This NASA news page carries more technical stories that are sometimes not available elsewhere.

oposite.stsci.edu/pubinfo/Latest.html The latest photographs and news releases from the Hubble Space Telescope are posted here.

www.gsoc.dir.de/satvis/ This excellent Website provides the times and sky locations of all Earth satellites passing over your site. Simply load your latitude and longitude, and the list appears.

www.skypub.com/ Sky Publishing's Website carries excellent weekly astronomy news updates as well as libraries of useful information for amateur astronomers and links to many other astronomy Websites.

As in all hobbies, there are Internet chat sites (the main "newsgroup" on Usenet is: sci.astro.amateur), where like-minded people exchange views and ask or answer questions. Much of what appears is sound information, but keep in mind that opinion and fact—especially with regard to commercial equipment—can become intertwined in these open forums, where individuals are free to post anything they like.

Amateur Astronomy Clubs

One of the best decisions that I ever made in connection with amateur astronomy was to join an astronomy club. In my case, it was the Toronto Centre of The Royal Astronomical Society of Canada (RASC), 136 Dupont Street, Toronto, Ontario M5R 1V2 (888-924-7272 or 416-924-7973; Website: www.rasc.ca). The RASC has branches in all major Canadian cities. Its 3,000 members receive several useful publications, including the indispensable *Observer's Handbook*. In the United States, there is no national astronomical society comparable to the RASC, although the Astronomical Society of the Pacific (390 Ashton Avenue, San Francisco, California 94112; 415-337-1100), an 8,000-strong group of amateur and professional astronomers in the western states, is a regional organization similar to the RASC.

Almost every city in North America with a population of more than 50,000 has an astronomy club. Members can join eclipse expeditions or participate in group observing sessions, which are excellent opportunities to seek advice and to examine a variety of telescopes. Some astronomy groups operate a club observatory built from membership funds and donations. Membership may entitle you to a discount on magazine subscriptions and astronomy books.

There are several methods of locating the nearest astronomical society. The Websites of the major astronomy magazines (see page 170) have up-to-date listings. A good place to start is to call the stores listed in the Yellow Pages under "Telescopes," since such shops are in constant touch with amateur astronomers. Another approach is to contact the nearest science museum or planetarium. A staff member should know whether there is an active astronomy club and, if so, where it meets. Failing that, get in touch with the astronomy instructor at the local college. If these avenues prove fruitless, there may not be an astronomy club in your area, but as a last resort, call the library or regional newspaper, both of which are constantly involved with clubs and organizations of all types.

Astronomy Conventions

Each year, at more than 50 locations across North America, amateur astronomers gather for annual conventions to share their hobby, exchange ideas and look through telescopes. Experienced practitioners give talks and conduct workshops, a noted guest speaker entertains, telescope dealers and manufacturers display their product lines, and everyone has fun. Conventions have blossomed into a major element of the hobby. Some are very well organized, offering a unique opportunity to meet a wide range of astronomy aficionados. If you are new to backyard astronomy, I can think of no better way to experience instant full immersion.

Most astronomy conventions are held during the summer; all are listed

several months in advance in *Astronomy*, *Sky & Telescope* and *SkyNews* magazines, along with addresses and Websites for more information. Among the biggest and best annual conventions are: the Texas Star Party, held at a ranch in southwest Texas; Stellafane, near Springfield, Vermont; Riverside Telescope Makers Conference, Big Bear, California; Astrofest, central Illinois; Winter Star Party, Florida Keys (February); and Starfest, Mount Forest, Ontario. Most astronomy conventions are held over two or three days in a location where the skies are good for celestial viewing. If one is within a reasonable driving distance, don't miss it.

Observatories

Since a visit to a major astronomical observatory was a pivotal event in my infatuation with things cosmic, I strongly recommend such an excursion to any budding astronomy enthusiast.

Not all observatories have the time or the staff to offer visitors a peek through their telescopes. Some of the main observatories that do are: Lick Observatory, Mount Hamilton, California, a half-hour's drive east of San Jose (408-459-2513; Website: www.ucolick.org); Leander McCormick Observatory at the University of Virginia, Charlottesville, Virginia (804-924-7494; Website: www.astro.virginia.edu/); Allegheny Observatory, Pittsburgh, Pennsylvania (412-321-2400); Dearborn Observatory, Northwestern University, Evanston, Illinois (847-491-7650; Website: www.astro.nwu.edu/); McDonald Observatory, Fort Davis, Texas (915-426-3640; Website: www.as.utexas.edu/mcdonald/mcdonald.html); Lowell Observatory, Flagstaff, Arizona (520-774-2096; Website: www.lowell.edu/); Griffith Observatory, Los Angeles, California (323-664-1181 or -1191; Website: www.GriffithObs.org); Chabot Observatory, Oakland, California (510-530-3480; Website: www.cosc.org/); David Dunlap Observatory, Richmond Hill, Ontario (905-884-2112; Website: www.astro.utoronto.ca/ddo-home.html); Dominion Astrophysical Observatory, Victoria, British Columbia (250-363-0012 or -0001).

All these observatories have specific hours set aside for visitors, and some have elaborate exhibit areas. On cloudy nights, slide or movie programs may be available in addition to tours of the facility. Reservations may be required, so call for schedules well in advance.

The 200-inch Hale telescope at Palomar Observatory on Mount Palomar, a two-hour drive northeast of San Diego, California, is the largest telescope in the world open to tourists. Thirteen stories high and enclosed by a massive rotating dome, the instrument (left) is awesome. For research reasons, views of the 200-inch telescope are restricted to a glass-enclosed visitors' gallery and exhibit area and to daytime hours (Website: astro.caltech.edu/observatories/palomar/public/index.html).

Kitt Peak National Observatory, 45 minutes southwest of Tucson, Arizona, is spectacular. During the day, you can take a paved road right up the mountain, where you'll find exhibits and impressive views of massive telescopes at an exquisite site (520-318-8726; Website: www.noao.edu/outreach/kpoutreach.html).

The largest collection of major research observatories in the world is at the summit of Mauna Kea, an extinct volcanic mountain on the island of Hawaii. No visitors are permitted at the summit at night, but the observatories have a comfortable visitors' center two-thirds of the way up the mountain, where telescopes are available for viewing on many evenings throughout the year. Contact Mauna Kea Observatories, Onizuka Visitors Information Station, 177 Makaala Street, Hilo, Hawaii 96720 (808-961-2180; E-mail: mkvis@ifa.hawaii.edu).

Telescope Equipment & Accessories

The following list includes many of the leading manufacturers of astronomy equipment and brief summaries of their products. Most of these companies will supply a product catalog or brochure upon request. This is *not* a listing of astronomy-equipment *dealers*. Many dealers sell all brands, and a visit to one will give you an opportunity to see and compare products from many companies. A local telescope dealer can be a valuable interface between you and a distant manufacturer, especially if you need your equipment serviced. For the dealer closest to you, check the Yellow Pages of the nearest large city under "Telescopes" or contact the manufacturer for a dealer list. Many dealers also have advertisements in the major astronomy magazines.

ASTRO-PHYSICS, INC.
11250 Forest Hills Rd.
Rockford, IL 61115
815-282-1513; fax: 815-282-9847
Apochromatic refractors, high-precision equatorial mounts.

CELESTRON INTERNATIONAL
2835 Columbia St., Torrance, CA 90503
310-328-9560; fax: 310-212-5835
Website: www.celestron.com
Full line of refractors, Newtonians, Schmidt-Cassegrains, eyepieces and telescope accessories.

COULTER OPTICAL
1781 Primrose Lane
W. Palm Beach, FL 33414
561-795-2201; fax: 561-795-9889
Website: www.murni.com
Dobsonian-mounted Newtonians.

EDMUND SCIENTIFIC COMPANY
101 E. Gloucester Pike
Barrington, NJ 08007
609-573-6250; fax: 609-573-6295
Astroscan telescope, eyepieces, lenses, optical components.

JIM KENDRICK STUDIO
2775 Dundas St. W.
Toronto, ON M6P 1Y4, Canada
416-762-7946; fax: 416-762-2765
Website: kendrick-studio.com
Dew-prevention heaters and other accessories.

JIM'S MOBILE INC.
810 Quail St., Unit E
Lakewood, CO 80215
303-233-5353; fax: 303-233-5359
Telescope accessories.

LUMICON
2111 Research Dr., #5A
Livermore, CA 94550
925-447-9570; fax: 925-447-9589
Nebula filters and photo accessories.

MAG 1 INSTRUMENTS
16342 Coachlight Dr.
New Berlin, WI 53151
414-785-0926; fax: 414-821-2133
Website: www.mag1instruments.com
Portaball 8" and 12" Newtonians.

MEADE INSTRUMENTS CORPORATION
6001 Oak Canyon
Irvine, CA 92620
714-451-1450; fax: 714-451-1460
Website: www.meade.com
Telescopes of all types and sizes, plus extensive array of accessories.

OBSESSION TELESCOPES
Box 804, Lake Mills, WI 53551
920-648-2328; Website:www.globaldialog.com/~obsessiontscp/OBHP.html
Premium-quality Dobsonians.

ORION TELESCOPE & BINOCULAR CENTER
Box 1815
Santa Cruz, CA 95061
408-763-7000; fax: 408-763-7017
Website: www.oriontel.com
Wide range of introductory telescopes and binoculars.

QUESTAR CORP.
6204 Ingham Rd.
New Hope, PA 18938
215-862-5277; fax: 215-862-0512
Questar Maksutov-Cassegrains.

SKY INSTRUMENTS
Box 3164, Vancouver, BC V6B 3Y6
Canada; 604-270-2813
Eyepieces and introductory-level telescopes.

STARGAZER STEVE
1752 Rutherglen Cres., Sudbury,
ON P3A 2K3, Canada; 705-566-1314
Website: ww2.isys.ca/stargazer
Beginner-level telescopes and telescope kits.

STARSPLITTER TELESCOPES
3228 Rikkard Dr.
Thousand Oaks, CA 91362
805-492-0489; fax: 805-492-0489
Website: www.starsplitter.com
Premium-quality Dobsonians.

TAURUS TECHNOLOGIES
Box 14, Woodstown, NJ 08098
609-769-4509
Astrophotography equipment.

TECTRON TELESCOPES
3544 Oak Grove Dr., Sarasota,
FL 34243; 941-355-2423; Website:
www.icstars.com/tectron/index.html
Premium-quality Dobsonians.

THOUSAND OAKS OPTICAL
Box 4813
Thousand Oaks, CA 91359
805-491-3642; fax: 805-491-2393
Solar filters.

UNIVERSITY OPTICS, INC.
Box 1205
Ann Arbor, MI 48106
313-665-3575; fax: 313-665-1815
Website: www.universityoptics.com/
Eyepieces and accessories.

VIRGO ASTRONOMICS
608 Falconbridge Dr., #46
Joppa, MD 21085
410-679-7055; fax: 410-679-2376
E-mail: virgoastronomics@erols.com
Binocular supports.